Classroom Companion: Economics

The Classroom Companion series in Economics includes undergraduate and graduate textbooks alike. It welcomes fundamental textbooks aimed at introducing students to the core concepts, empirical methods, theories and tools of the field, as well as advanced textbooks written for students at the Master and PhD level seeking a deeper understanding of economic theory, mathematical tools and quantitative methods.

Giuseppe Travaglini · Giorgio Calcagnini ·
Alessandro Bellocchi

Notes on Consumption Theory

Deterministic and Stochastic Dynamic Models

Springer

Giuseppe Travaglini
Department of Economics, Society, and Politics
University of Urbino Carlo Bo
Urbino, Italy

Giorgio Calcagnini
University of Urbino Carlo Bo
Urbino, Italy

Alessandro Bellocchi
University of Urbino Carlo Bo
Urbino, Italy

ISSN 2662-2882 ISSN 2662-2890 (electronic)
Classroom Companion: Economics
ISBN 978-3-031-54985-4 ISBN 978-3-031-54986-1 (eBook)
https://doi.org/10.1007/978-3-031-54986-1

This Springer imprint is published by the registered company Springer Nature Switzerland AG
The registered company address is: Gewerbestrasse 11, 6330 Cham, Switzerland

Paper in this product is recyclable.

"Consumption is the sole end and purpose of production".

Adam Smith
(Wealth of Nations, 1776, Book IV, Ch. 8)

"Consumption may be regarded as negative production".

Alfred Marshall
(Principles of Economics, 1890, Book II, Ch. 3)

"All production is for the purpose of ultimately satisfying a consumer".

John Maynard Keynes
(The General Theory of Employment, Interest and Money, 1936, Ch. 5)

"[...] much less emphasis should be attached to the regressions of consumption on income".

Milton Friedman
(A Theory of the Consumption Function, 1957, Ch. 9)

To our students

Preface

This book presents the teaching material "Notes on Consumption Theory. Deterministic and Stochastic Dynamic Models" written for doctoral courses at the University of Urbino Carlo Bo. Also part of the collection are notes drafted over the course of at least a decade for other Ph.D. classes, such as those held at Marche Polytechnic University in Ancona and La Sapienza in Rome, and notes prepared for lectures addressed to Urbino students in the Master's degree courses. Some sections also refer to articles we have published in academic journals.

The aim is to bring these contributions together in a coherent analytical corpus to provide graduate students with a concise book on intertemporal consumption theory. For the sake of clarity, we begin with a two-period model. We then generalize the setting to the T-period model with an infinite horizon, focusing on the model's dynamic properties in deterministic and stochastic scenarios.

Once we have (hopefully) clarified that the intertemporal optimal choice must satisfy a mathematical condition known as the *Euler equation*, any subsequent generalization of the model to increasingly complex cases becomes relatively simple.

From an analytical point of view, the difference and differential equations are keys to the analysis. We present optimization techniques, including the Lagrangian and Hamiltonian functions, the maximum principle and the dynamic programming with the Hamiltonian-Jacobi-Bellman equation. Further, Markovian processes, Brownian motions, stochastic dynamic programming, Ito's lemma, and stochastic optimal control are tools that we use extensively to capture the complexity of the intertemporal consumption models under conditions of uncertainty, in discrete and continuous time. Importantly, we focus on the relationship among consumption, saving, investment (real and financial), and economic growth. A section of the book is indeed devoted to the Ramsey-neoclassical growth model, with and without technological progress, formalizing its extensions to the issues of sustainable and stochastic consumption.

This book is, in essence, an introduction to the broad issues of dynamic consumption theory. Therefore, it is primarily aimed at Ph.D. students in economics and finance, MBA students specializing in the fields of finance and stochastic quantitative methods, and researchers concerned with decision theory under conditions of certainty and uncertainty who may find in its pages useful materials for research and advanced teaching. Moreover, the topics developed here may be useful to Master's students in the field of economics and quantitative analysis who are curious about dynamic mathematics and its application to intertemporal economic problems, as is the case for consumption theory.

A final warning. Consumption theory has many applications in theoretical and empirical fields. These notes emerge from our selection of topics that emphasize the intertemporal aspects of economic decisions. Thus, some aspects, such as capital market imperfections or stochastic continuous-time models, are treated formally, although we maintain an introductory approach. Further, readings and resources for additional study are listed at the end of each chapter.

This book owes a great deal to many people. We want to thank the many students in our courses for their feedback, patience, and encouragement. Two anonymous referees provided detailed, thoughtful, and constructive comments on many aspects of the book. Each significantly improved the end result, and we are deeply grateful for their efforts.

Many others have made valuable comments and suggestions concerning some or all of the book's contents. Special thanks go to Enrico Saltari, Emeritus Professor in Economics, whose teachings, both in economic theory and mathematical analysis, form the foundation of our knowledge and research developed over the years. We would particularly like to thank Andrea Bacchiocchi, Gian Italo Bischi, Edgar Sanchez Carrera, Andrea Coveri, Federico Favaretto, Laura Gardini, Alessandra Giannantonio, Germana Giombini, Paolo Liberati, Giovanni Marin, Agnese Sacchi, Davide Ticchi, Fabio Tramontana, and Antonello Zanfei. Finally, the editorial and production staff at Springer Nature did an excellent job of turning the final draft of the book into a finished product. We thank everyone for their efforts.

Urbino, Italy
2024

Giuseppe Travaglini
Giorgio Calcagnini
Alessandro Bellocchi

Contents

About the Authors

Giuseppe Travaglini Professor of Economics at the University of Urbino Carlo Bo. He is currently Head of the Department of Economics, Society, Politics (DESP). He teaches Macroeconomics and Financial Markets. His research interests include consumption theory, investment, dynamic models, technical change, and economic growth.

Giorgio Calcagnini Professor of Economics at the University of Urbino Carlo Bo. He is currently the Rector of the same University. He teaches Econometrics and Monetary Economics. His research focuses on the role of market imperfections, firms' investment decisions, structural change in the banking industry, and financial literacy and its relation with consumption.

Alessandro Bellocchi He is a Postdoctoral Research Fellow at the Department of Economics, Society, Politics (DESP) of the University of Urbino Carlo Bo. He currently teaches European Economic Policy. His research interests include income distribution, labor market, consumption theory, technological change and economic growth.

1 Consumption and Certainty

- In this chapter, we study *intertemporal consumption* choices under conditions of certainty.
- We use a *two-period model* where an individual chooses between current and future consumption, given the *intertemporal budget constraint.*
- The saving decision becomes consumption in the next period, thus investment and future production.
- We explain the significance of the optimal condition, known as the *Euler equation* and consider the role of the *interest rate* in consumption decisions.
- Next, we extend the model to *multiple periods*, up to an infinite-time horizon.
- We introduce *liquidity constraints* to show how capital market imperfections affect consumption choices.
- The properties of *discrete-time* models are compared with those in *continuous time*.

1.1 Introduction

In this chapter, we present the basic elements of *consumption theory*. The process regulating intertemporal consumption is fundamental to understand the link between income and consumption, how transitory and permanent income affect saving decisions, how saving and investment are related, and how consumption decisions impact financial markets and individual portfolios.

Analytical representation of these processes requires simplified models that are useful to study a single element or aspect among the many that characterize consumption decisions. Although this approach may limit the realism of the model, it allows us to disentangle the role of the various factors affecting the consumer's decision process.

G. Travaglini et al., *Notes on Consumption Theory*, Classroom Companion: Economics,
https://doi.org/10.1007/978-3-031-54986-1_1

For this reason, we begin with the analysis of a *two-period model* under conditions of *certainty*. As we shall see, this basic framework includes features of more complex models—consumption models with a longer planning horizon, in discrete or continuous-time, deterministic or stochastic.

Notably, the initial assumption of *no uncertainty* allows deeper insight into the economic significance of the *Euler equation*, which, as we point out, drives optimal consumption choices over time. This scenario also allows us to focus more easily on the significance of the *transversality condition*, analyze the role of permanent income in the consumption function, and study the evolution of savings and wealth accumulation with its lasting effects on financial markets, sustainable consumption, and portfolio choices.

1.2 A Two-Period Model

Consider a consumer (individual) who lives for two periods and must decide to allocate his or her consumption between t and $t + 1$. Decisions are made in the absence of uncertainty. The intertemporal consumption choice is the result of *utility maximization* by the consumer. No inheritance affects the individual's consumption plan over the two periods.

At time $t = 0$, the individual has an initial endowment equal to a_0 and receives labor income y_0. Income in $t = 1$ is known with certainty and is equal to y_1. Finally, we assume the possibility of borrowing and lending in a perfect capital market.

Since an individual usually does not allocate his or her entire endowment to current consumption, we denote saving in the present time by $(a_0 + y_0) - c_0$, c_0 being the consumption for the period. Finally, r is the interest rate either received on savings or paid on loans.

Thus, wealth a_t, in different periods, is as follows:

$$t = 0 : a_0 \tag{1.1}$$

$$t = 1 : a_1 = (1 + r)(a_0 + y_0 - c_0) \tag{1.2}$$

$$t = 2 : a_2 = (1 + r)(a_1 + y_1 - c_1) \tag{1.3}$$

Equations. 1.1–1.3 represent the total wealth available in each period. However, since we assume that the consumer lives for two periods only and that there are no inheritances, the final wealth in $t = 2$ must be $a_2 \leq 0$. In fact, it is not an optimal choice to have positive wealth in the period $t = 1$ (axiom of non-satiety). In other words, we are assuming that each individual desires to maximize his or her intertemporal consumption and that there is no intergenerational solidarity.

Further, to rule out the possibility of debts remaining at the end of the period, the condition $a_2 \geq 0$ (positive final wealth) must also be verified. Indeed, it is not possible to raise consumption with an increasing degree of indebtedness.

These two conditions together imply that

$$a_2 = 0, \tag{1.4}$$

which we refer to as the *transversality condition*. Substituting Eq. 1.4 in Eq. 1.3 in $t = 2$, we have

$$a_1 = c_1 - y_1. \tag{1.5}$$

Then, substituting Eq. 1.5 backward for a_1 in Eq. 1.2 in $t = 1$, we obtain the following:

$$c_1 - y_1 = (1 + r)(a_0 + y_0 - c_0). \tag{1.6}$$

Equation 1.6 is simply the *intertemporal budget constraint*

$$c_0 + \frac{c_1}{1 + r} = a_0 + y_0 + \frac{y_1}{1 + r} \tag{1.7}$$

The relationship in Eq. 1.7 states that the *present value* of consumption (left-hand side) cannot be different from the present value of total wealth, that is, the sum of the initial endowment, a_0, plus the discounted value of income received in the two periods (right-hand side). This means that if, as in our example, the price of current consumption is equal to 1, the price of discounted future consumption is $\frac{1}{1+r}$, or the number of units of future consumption that the consumer must give up to increase current consumption by one unit.

Figure 1.1 depicts the intertemporal budget constraint in Eq. 1.7 under the simplifying assumption that $a_0 = 0$. The straight line is decreasing with a slope, in absolute value, equals to the interest factor $(1 + r)$. The point D identifies one of the possible endowments in the two periods. The two intercepts along the x- and y-axes are equal to $y_0 + y_1/(1 + r)$ and $y_0(1 + r) + y_1$, respectively. If the consumer spends all of his or her endowment in the second period, then $c_0 = 0$ and $c_1 = (1 + r)y_0 + y_1$. If, however, the consumer is impatient and anticipates consumption in the current period, $c_1 = 0$ and $c_0 = y_0 + y_1/(1 + r)$.[1] To sum up, the budget constraint identifies all feasible intertemporal combinations of consumption given the present value of the income endowment.

1.2.1 Optimal Allocation

Let $u(c_0, c_1)$ be the utility function representing a consumer's preferences between current and future consumption. We assume that utility is increasing in c_0, c_1, but

[1] As shown by Reuben et al. (2010) discount rates are linked to individual traits.

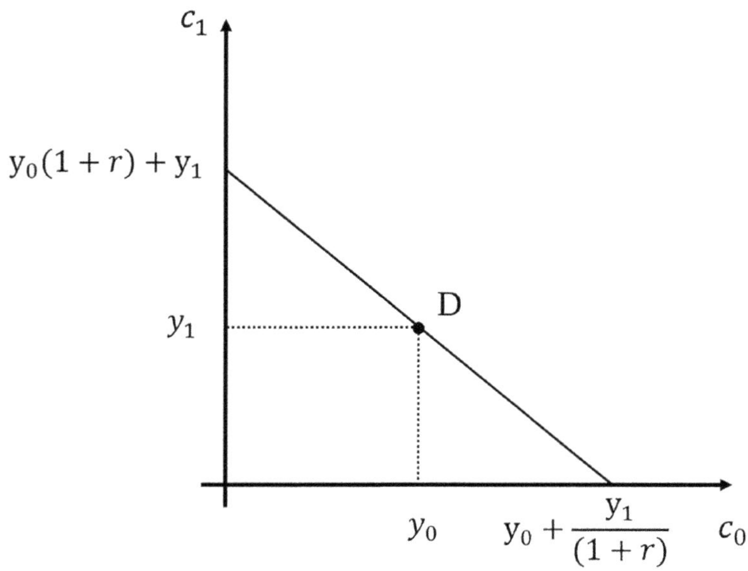

Fig. 1.1 The intertemporal budget constraint

with decreasing marginal utility, that is $\frac{\partial u}{\partial c_i} > 0$ and $\frac{\partial^2 u}{\partial c_i{}^2} < 0$. The consumer problem can then be written as

$$\begin{cases} \max\limits_{c_0, c_1} & u(c_0, c_1) \\ \text{s.t.} & c_0 + \frac{c_1}{1+r} = a_0 + y_0 + \frac{y_1}{1+r} \equiv a_0 + h_0, \end{cases} \tag{1.8}$$

with $h_0 \equiv y_0 + \frac{y_1}{1+r}$ and $c_0, c_1 > 0$. The corresponding Lagrangian function is

$$L = u\left(c_0, c_1\right) + \lambda\left(a_0 + h_0 - c_0 - \frac{c_1}{1+r}\right). \tag{1.9}$$

The optimal condition requires that the marginal rate of intertemporal substitution (MRS) is equal to the price ratio:

$$\frac{u_{c0}}{u_{c1}} = (1+r), \tag{1.10}$$

where $u_{c0} = \frac{\partial u}{\partial c_0}$ and $u_{c_1} = \frac{\partial u}{\partial c_1}$. It follows that

$$u_{c_0} = (1+r)\, u_{c_1}. \tag{1.11}$$

This relation is known as the *Euler equation*. At the tangency point, which identifies the optimal solution for present and future consumption, the slope of the indifference curve is equal to the one of the constraint (Fig. 1.2). Equation 1.11 states that

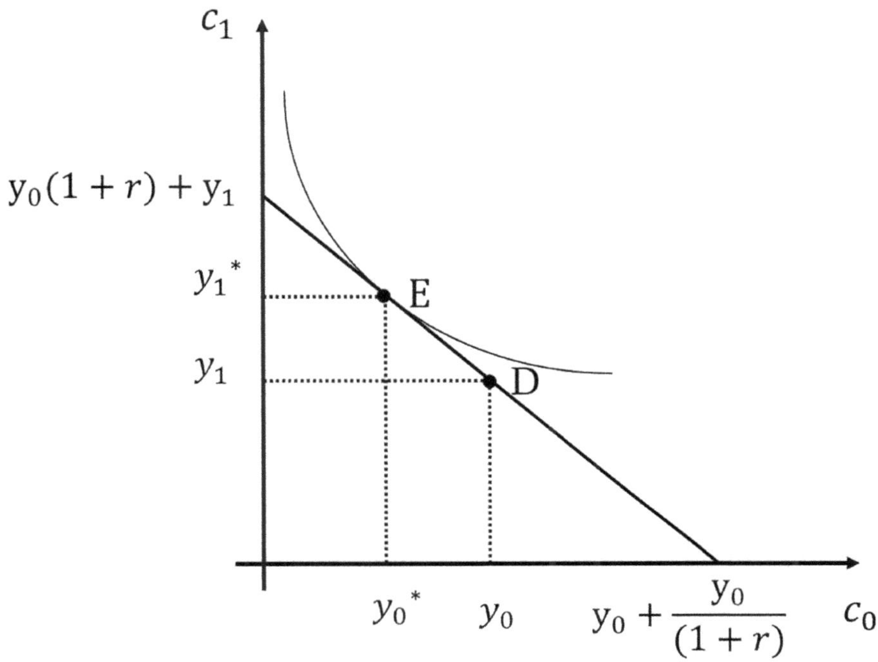

Fig. 1.2 The intertemporal optimum

the intertemporal consumption choice is optimal when the marginal value of consumption between two consecutive periods is equal to the interest factor. In fact, a decrease dc_0 in current consumption increases the individual's marginal utility at $t = 0$, and decreases future marginal utility by $(1 + r)\, u_{c_1}$, since in $t = 1$, consumption increases by $(1 + r)\, dc_0 > 0$. The Euler equation thus provides the optimal path of marginal utility in each period.

Let us apply what we have learned thus far to some utility functions, namely the *additively separable* and *Cobb-Douglas* utility functions.

Example 1.1 (*Additively separable utility function*) Suppose that the utility function $u\,(c_0, c_1)$ has the following form:

$$u\,(c_0, c_1) = u\,(c_0) + \frac{1}{1+\delta} u\,(c_1)\,, \tag{1.12}$$

where δ measures the degree of the consumer's intertemporal preference (i.e., his or her level of impatience). The first-order conditions are as follows:

$$u_{c_0} = \lambda, \tag{1.13}$$

$$u_{c_1} = \lambda \frac{1+\delta}{1+r}. \tag{1.14}$$

Combining these two relations, we get the Euler equation:

$$\frac{u_{c_0}(1+\delta)}{u_{c_1}} = 1 + r. \tag{1.15}$$

Note that if $u(c_t) = \ln c_t$, the previous condition becomes

$$c_1 = \left(\frac{1+r}{1+\delta}\right) c_0. \tag{1.16}$$

It follows that

- *If* $r = \delta$, optimal consumption is *constant* over time ($c_1 = c_0$);
- *If* $r > \delta$, then $c_1 > c_0$, implying that optimal consumption *grows* over time. In fact, if the return on savings is higher than the rate of intertemporal preference δ, the consumer prefers to postpone consumption to a future period.
- *If* $r < \delta$, then $c_1 < c_0$, implying that optimal consumption *decreases* over time. This happens because the interest rate r is not sufficiently high to compensate for the impatient consumption.

Example 1.2 (*Cobb-Douglas utility function*) Suppose that the utility function $u(c_0, c_1)$ is as

$$u(c_0, c_1) = c_0 c_1. \tag{1.17}$$

Further, assume that $a_0 = 0$ and that the individual's income in the two periods is $y_0 = 10$ and $y_1 = 30$. The market interest rate is $r = 0.02$, and the budget constraint is given by the expression in Eq. 1.7.

The optimal condition can easily be rewritten as follows:

$$\frac{c_1}{c_0} = 1.02. \tag{1.18}$$

We solve the system for c_0 and c_1, obtaining

$$\frac{c_1}{c_0} = 1.02, \tag{1.19}$$

$$c_0 + \frac{c_1}{1.02} = 10 + \frac{30}{1.02}. \tag{1.20}$$

The solutions are $c_0 = 19.7$ and $c_1 = 20.1$. Therefore, the consumer finds it optimal in the first period to borrow ($c_0 > y_0$), obtaining an initial credit equal to 9.7. This credit becomes a debt in the second period and is increased by the interest rate; that is, $(c_0 - y_0)(1 + r) = 9.7(1.02) = 9.9$. This explains why, in $t = 0$, consumption is $c_1 = y_1 - (c_0 - y_0)(1 + r) = 20.1$ ($c_1 < y_1$).

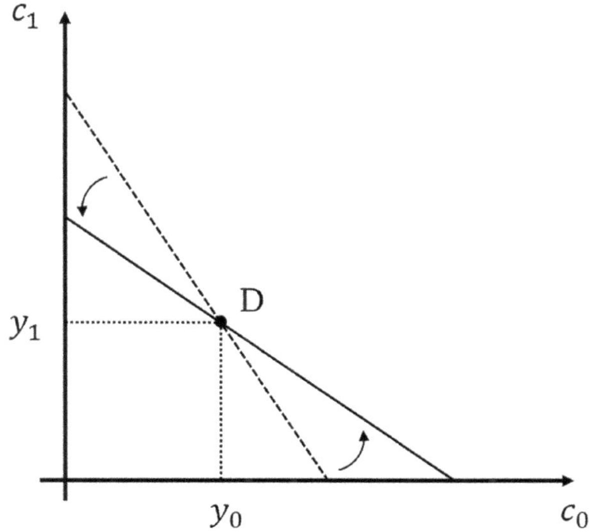

Fig. 1.3 The effects of a reduction of r on the budget constraint

1.2.2 Changes in the Interest Rate

Next, we study how the consumer's optimal choice varies in response to changes in the interest rate r. A change in r has a direct effect on the intertemporal budget constraint in that it affects the price $\frac{1}{1+r}$ of future consumption c_1 and the slope of the straight line representing it. If, for example, r decreases, the budget constraint becomes less *steep*, rotating counterclockwise around the point D (Fig. 1.3). The present value $y_0 + y_1/(1+r)$ increases, while the value of income $(1+r)y_0 + y_1$ decreases. Hence, the intercept on the x-axis shifts to the right, while the intercept on the y-axis shifts downward.

We study the effect of a change in the interest rate on consumption and saving by focusing on the *income* and *substitution* effects. Suppose (for convenience) that an individual is a saver; that is, $c_0 < y_0$. In this case, if r decreases, the following occur simultaneously:

- *Substitution effect*: a decrease in r increases current consumption, thus *decreasing savings*.
- *Income effect*: a decrease in r decreases the return on savings. If current consumption is a *normal good*, the decrease in income eventually implies a decrease in current consumption, thus *increasing savings*.

Therefore, since the income and substitution effects operate in opposite directions, the overall impact on current consumption is ambiguous (i.e., in Fig. 1.4a the saving does not change).

But, what if the individual is impatient and anticipates future consumption by *borrowing*? Now, the following effects occur simultaneously:

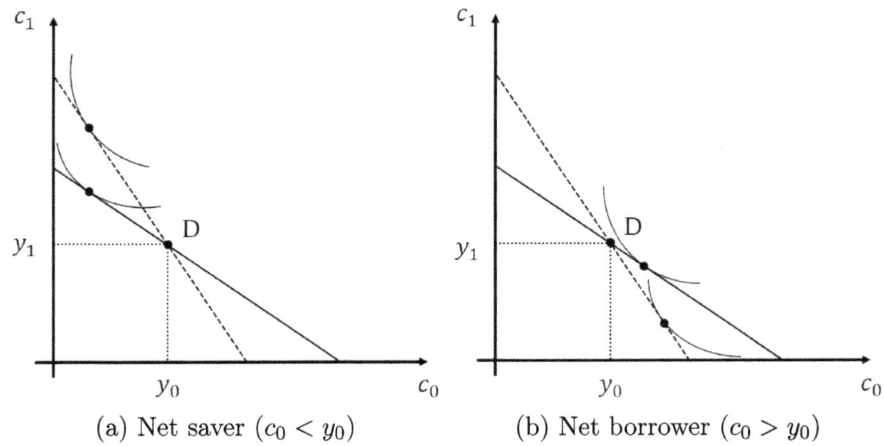

Fig. 1.4 The interest rate and consumption choices

- *Substitution effect*: a decrease in r causes an increase in first-period consumption, so saving necessary decreases.
- *Income effect*: the decrease in r makes borrowing cheaper, thus encouraging the consumer to further increase his or her demand for loans.

Hence, if the individual is initially borrowing the income effect goes in the same direction as the substitution effect, both reducing c_0 and thus increasing the savings in the initial period (Fig. 1.4b).

Finally, note that since the stock of wealth in the economy is positive, individuals are on average savers rather than borrowers. Thus, the income effect of a rise (decrease) in the interest rate is positive (negative).

1.3 Consumption and Investment

Saving and investment are two sides of the same coin, and their relationship deserves special attention. A basic model is provided by Irving Fisher (1930) in *The theory of interest*. He shows that with *perfect capital markets*, consumption and investment decisions are *separate*.

We explore the point by focusing on an individual who must choose how to allocate his endowment over two periods. However, in contrast to the previous exercise, he or she faces the *dual problem* of consuming and producing. Therefore, alongside the utility function, we introduce a production function to transform savings into future production (and hence consumption).

Before continuing, we note that Fisher's model is a *general* equilibrium model. It is not the first to deal with the topic; in an earlier study Ramsey (1928) provides a framework of intertemporal resource allocation with an infinite horizon. His famous article on optimal saving behavior has had an enormous, albeit delayed, influence on the current literature on optimal consumption and economic growth. Today, Ramsey's model is at the heart of the neoclassical model of economic growth. We will study this model in the second part of the book.

1.3.1 Technology Constraints

Let us start with the *production problem*. We assume that any individual is simultaneously a firm and a consumer, and, as in the previous case, he or she lives for two periods only. In this setup, c_t is the dividend, measured in units of consumption, that is consumer's income in each period.

Since income depends on output, it is possible to derive a relation between consumption in $t = 0$ and $t = 1$ by varying the use of inputs in each period, that is, by investing (or disinvesting).

We assume that the endowment K_0 is a datum of the problem. Therefore, the investment at $t = 0$ can be written as $I_0 = K_1 - K_0$, where K_1 is the stock at $t = 1$. Finally, the production function $F(K_t) = K_t^\alpha$ has decreasing marginal returns since $\alpha < 1$. The price of input is equal to 1.

In $t = 0$, the output is given since it depends on the endowment K_0. It can be allocated between dividends (consumption) and investment so that

$$F(K_0) = c_0 + I_0. \tag{1.21}$$

Since the consumer will exhaust all available production before he dies, an analogous condition must hold even in the following period, as follows:

$$F(K_1) = c_1. \tag{1.22}$$

Using the production function $F(K_t) = K_t^\alpha$, Eqs. 1.21 and 1.22 can be rewritten as

$$K_0^\alpha = c_0 + (K_1 - K_0), \tag{1.23}$$

$$K_1^\alpha = c_1 \quad \text{with} \quad I_1 = 0. \tag{1.24}$$

Then, by combining Eqs. 1.23 and 1.24 we get

$$\left(K_0^\alpha + K_0 - c_0\right)^\alpha - c_1 = 0 \equiv T\,(c_0, c_1)\,. \tag{1.25}$$

where $T\,(c_0, c_1)$ is the frontier of production possibilities or, the *transformation curve*.

This curve is represented in Fig. 1.5. It describes the feasible combinations of output obtained by various investment decisions, that is, by changing K_1 along the curve $T\,(c_0, c_1)$. So, for each level of current consumption, this curve sets the corresponding maximum level of output available in the next period.

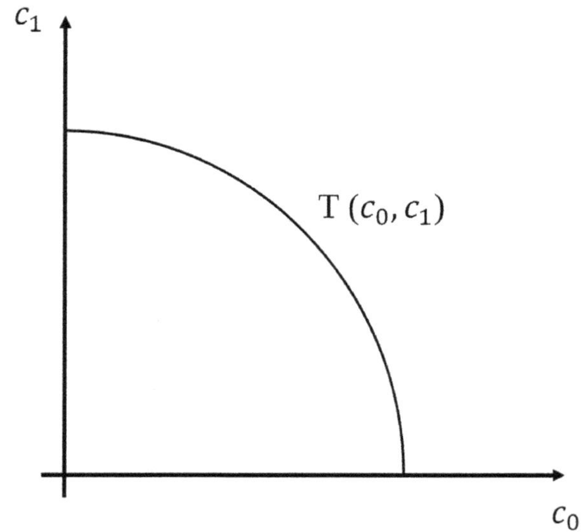

Fig. 1.5 The transformation curve

1.3.2 No Capital Markets

Following Fisher (1930), we assume that initially capital markets *do not exist*. The maximization problem can be written as

$$\begin{cases} \max\limits_{c_0, c_1} u(c_0, c_1) \\ \text{s.t.} \quad T(c_0, c_1) = 0. \end{cases} \tag{1.26}$$

The optimum is reached when the marginal rate of substitution (MRS) equals the marginal rate of transformation (MRT)

$$MRS = MRT; \tag{1.27}$$

that is, when

$$\frac{u_{c_0}}{u_{c_1}} = \alpha \left(K_0^\alpha + K_0 - c_0\right)^{\alpha-1} = \alpha (K_1)^{\alpha-1} \tag{1.28}$$

This result has a crucial implication: in the absence of capital markets, intertemporal consumption cannot be different from intertemporal production. In other words, consumption and investment decisions *coincide* because the equilibrium point E is where the transformation curve is tangent to the indifference curve. Indeed, since K_0 is the endowment in $t = 0$, the corresponding output is $F(K_0)$. So, if c_0^* is consumption in the first period, the optimal level of savings is $K_1^* = F(K_0) - c_0^*$. In this scenario, consumption and investment decisions cannot be made independently of each other (Fig. 1.6).

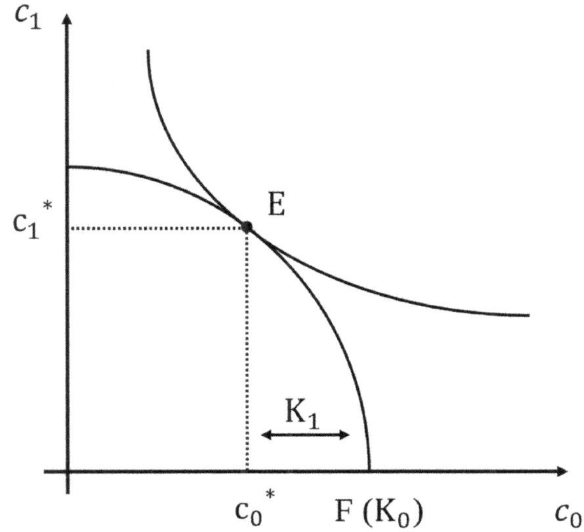

Fig. 1.6 Equilibrium without capital markets

1.3.3 Capital Markets

Assume now that capital markets exist. Consumers can borrow and lend at the interest rate r, so that consumption can differ from output in each period.

To clarify, in the real world, savings decisions are separate from investment decisions: consumers do not decide how much to produce, nor do firms produce with the goal of maximizing their utility. It is the market that connects individual preferences and production by means of demand and supply.

In our example, the production input can be consumed or invested. However, because production and consumption may require different distributions of resources over time, the exchange of resources takes place through specific markets. We define two capital markets that differ based on when the exchange is made:

1. *the spot market*, where the exchange takes place in the current period;
2. *the futures market*, where the exchange takes place in the next period.

These markets are complete and competitive. Finally, in this scenario, any individual can move both along the transformation curve and along the capital market line by lending and borrowing.

Note that, in equilibrium, the interest rate r in capital markets and the return π in production must be the same. Indeed, if π were *higher* than r, it would be worth taking on debt at the cost r, investing in production to obtain the extra return $\pi - r > 0$. This *arbitrage* opportunity cannot last forever. In fact, under the assumption of perfect markets, the demand for loans tends to increase, raising the rate r, and the real return π tends to decrease due to the assumption of decreasing marginal productivity. This process stops when $1 + r = 1 + \pi$. Hence,

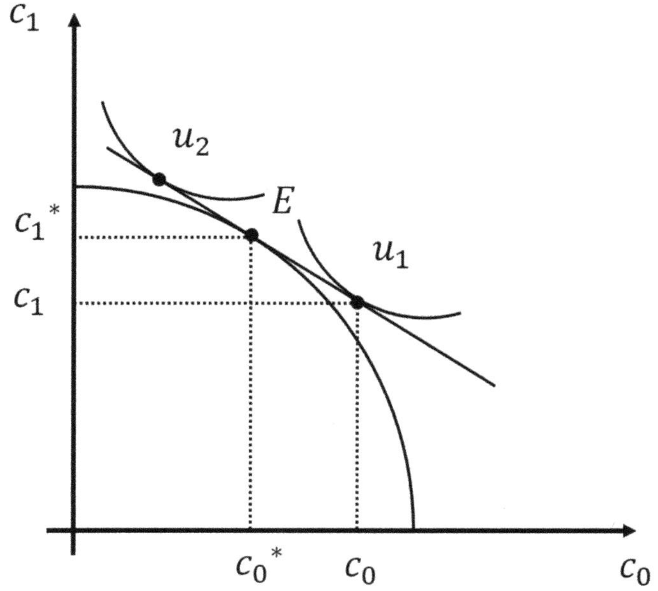

Fig. 1.7 Equilibrium with capital markets

$$\left|\frac{-\partial T/\partial c_0}{\partial T/\partial c_1}\right| = MRT, \tag{1.29}$$

or

$$1 + r = \alpha \left(K_0^\alpha + K_0 - c_0\right)^{\alpha-1} \tag{1.30}$$

$$1 + r = \alpha (K_1)^{\alpha-1}. \tag{1.31}$$

It follows that, in $t = 0$,

$$c_0^* = K_0^\alpha + K_0 - \left(\frac{1+r}{\alpha}\right)^{\frac{1}{\alpha-1}}. \tag{1.32}$$

In $t = 1$

$$c_1^* = K_1^\alpha = \left(\frac{1+r}{\alpha}\right)^{\frac{\alpha}{\alpha-1}}. \tag{1.33}$$

Figure 1.7 depicts the consumer's decision along the curves. Once the optimal production point E is reached, the individual will decide whether to lend or borrow based on his or her intertemporal preference rate, δ. The point u_1 represents an individual whose impatience leads to borrow in the current period. Alternatively, for an individual who attributes greater importance to future consumption, it is preferable to lend at $t = 0$, as in point u_2. As we know, it all depends on the relative magnitude of δ and r. Similarly, the allocation can be optimal even for a consumer with δ equal to r.

Therefore, we have now three different solutions: one involving *borrowing*, one involving *lending*, and one involving *neither*. It is important to point out that in each case, the solution to the optimal investment decision is *independent* of the consumption preference, which determines the consumption decision. Thus, there is a separation between the investment and consumption choices because of the assumption of a perfect capital market.

In conclusion, the investment decision only requires information on both the investment opportunities and the interest rate. The consumption decision only requires information on the utility function and the interest rate r.

In summary, with perfect capital markets, it is possible to separate production and consumption decisions. This requires, first, solving the investment problem and then solving the consumption problem. Finally, it is important to note that the separation principle implies that individuals can improve his or her welfare through capital markets. Indeed, the point of optimum for the borrowing or lending consumer is always on a higher indifference curve than it would be along the transformation curve. In short, the existence of a perfect capital market facilitates the intertemporal allocation of resources and improves individual welfare.

1.4 Imperfect Capital Markets

How does this result change under the assumption of *imperfect* capital markets? As an example, Fig. 1.8 shows what happens if the borrowing rate (r_D) is higher than the lending rate (r_A). This may be due to a variety of factors causing disparities between borrowers and lenders in terms of their relative risk, access to information, and market power (Calcagnini and Saltari 2009). Unlike the case of perfect markets, the optimal investment choice now depends on consumption preferences. If, for example, some individual prefer to lend current income (curve u_1) and others prefer to anticipate future consumption (curve u_2), there would be considerable difficulty in reconciling these goals. Indeed, the optimal investment would be realized at two different interest rates, resulting in a conflict between the investment and consumption decisions.

Capital-market imperfections take different forms and have different origins. They can result in price alterations or limited quantities. In the case presented above, lending and borrowing rates could differ. In other cases, consumers may suffer from *credit rationing*, a situation in which they cannot access the desired credit. When the consumer is "rationed", his or her consumption choices are limited by liquidity constraints. Let us discuss an example.

1.4.1 Credit Rationing

Consider an individual who can only borrow a small part of the amount desired. We say that he or she is *liquidity constrained*. The additional constraint affects the choice between consumption and saving, period by period, *and* the evolution of

Fig. 1.8 Imperfect capital markets

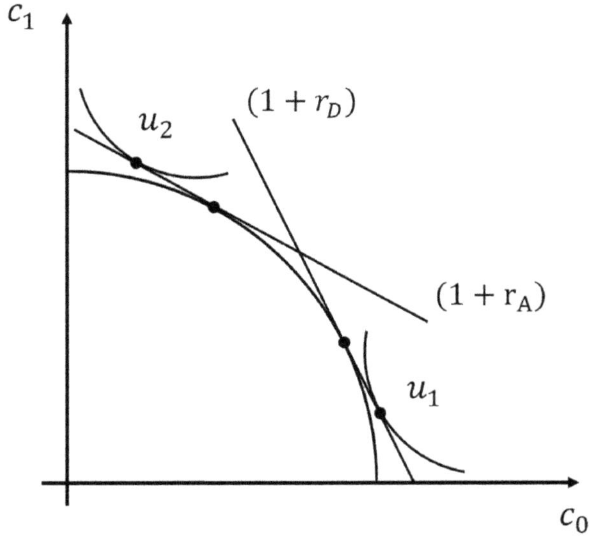

the individual's wealth (Hurst and Lusardi 2004). It is possible that, for example, individual wealth cannot fall below a specific threshold d_0 (debt), as determined by the willingness of external lenders, such as a bank or other financial institutions, to offer loans. We can write such a constraint as

$$a_0 \geqslant -d_0. \tag{1.34}$$

If $d_0 = 0$, the consumer cannot borrow, and his or her wealth will be $a_0 \geqslant 0$ (*complete* credit rationing): it is only his or her current and future income that determine consumption choices. If, on the other hand, $d_0 > 0$, the consumer can borrow up to, but not beyond, a certain amount (*partial* credit rationing). Note that the loans to which a consumer has access may also depend on his or her current income y_0. In fact, while an individual's assets can serve as collateral in case of default, it is income that banks use to assess the borrower's ability to make regular payments. In this case, we can write the constraint as $a_0 \geqslant -d(y_0)$, meaning that the amount borrowed depends on the consumer's personal income, which is used by the lender as a signal of his or her ability to repay the loan, including interest.

We assess the impact of liquidity constraints on intertemporal consumption decisions by returning to the two-period model. We assume initial wealth of zero and $a_1 \geqslant 0$; that is, the consumer can save but not get into debt (Zeldes 1989). Since $a_1 = y_0 - c_0 \geqslant 0$ the liquidity constraint also implies that $y_0 \geqslant c_0$. Therefore, consumption maximization now meets two different constraints: the *intertemporal budget* and the *liquidity* constraints.

The Lagrangian becomes

$$L = u(c_0, c_1) + \lambda \left(y_0 + \frac{y_1}{1+r} - c_0 - \frac{c_1}{1+r} \right) + \mu (y_0 - c_0). \tag{1.35}$$

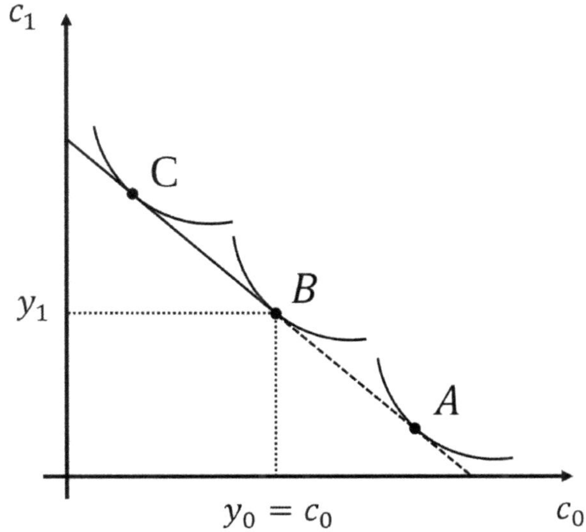

Fig. 1.9 Liquidity constraint and consumption

In Eq. 1.35 μ is the multiplier associated with the *liquidity constraint*. Now, the consumer's problem is to maximize his or her utility $u(c_0, c_1)$ while respecting both the budget constraint $c_0 + \frac{c_1}{1+r} = y_0 + \frac{y_1}{1+r}$ and the liquidity constraint $a_1 \geqslant 0$.

If the liquidity constraint binds, then $y_0 = c_0, a_1 = 0$ and $\mu > 0$. In this scenario, consumption is limited at time $t = 0$, and its level is given by the corner solution at the point where income and current consumption coincide since the desired consumption c^* is equal to or even greater than income y_0.

If, however, $\mu = 0$, then $y_0 > c_0$ and $a_1 > 0$; that is, the liquidity constraint does not bind (it is *slack*). In this case, the constraint does not affect the optimal choice because the desired consumption, c^* in $t = 0$, is less than y_0; the consumer saves today to consume tomorrow.

Figure 1.9 depicts the various scenarios described above. In the absence of liquidity constraints, the consumer can explore all combinations of consumption consistent with his or her intertemporal budget constraint. For example, given the endowment (y_0, y_1), point A describes the case in which the desired consumption c^* at time $t = 0$ is greater than current income, y_0. However, if liquidity constraints bind consumption in $t = 0$ such that it cannot exceed current income y_0, the consumer will be in the corner solution B, where $y_0 = c_0$. Note that, even in the case of credit rationing, if $c^* \leqslant y_0$, as at point C, the liquidity constraint $a_1 \geqslant 0$ has no effect (i.e., it does not bind) and the consumer chooses his or her intertemporal consumption *as if* he or she was not rationed.

This problem can be formalized with the Kuhn-Tucker conditions (Karush 1939; Kuhn and Tucker 1951) which generalize the Lagrange's method by including in the intertemporal model the inequality constraints $c_0, c_1 \geqslant 0$ and $a_1 \geqslant 0$ (see Lambert

1985; Chiang and Wainwright 2005). For the problem in Eq. 1.35, we obtain the following:

$$c_0 \geqslant 0; \quad \frac{\partial L}{\partial c_0} = u'(c_0) - \lambda - \mu \leqslant 0; \quad c_0\left(\frac{\partial L}{\partial c_0}\right) = 0. \tag{1.36}$$

$$c_1 \geqslant 0; \quad \frac{\partial L}{\partial c_1} = u'(c_1) - \frac{\lambda}{1+r} - \mu \leqslant 0; \quad c_1\left(\frac{\partial L}{\partial c_1}\right) = 0. \tag{1.37}$$

$$\mu \geqslant 0; \quad \frac{\partial L}{\partial \mu} = y_0 - c_0 \geqslant 0; \quad \mu\left(\frac{\partial L}{\partial \mu}\right) = 0. \tag{1.38}$$

$$\lambda \geqslant 0; \quad \frac{\partial L}{\partial \lambda} = y_0 + \frac{y_1}{1+r} - c_0 - \frac{c_1}{1+r} \geqslant 0; \quad \lambda\left(\frac{\partial L}{\partial \lambda}\right) = 0. \tag{1.39}$$

It is convenient to begin our study of the optimal conditions (Eqs. 1.36–1.39) with the hypothesis that the constraints are *binding*. If $\mu > 0$, the liquidity constraint binds and $y_0 - c_0 = 0$, implying that $\mu\frac{\partial L}{\partial \mu} = 0$. Similar reasoning applies to λ. Combining Eqs. 1.36 and 1.37, we obtain the following:

$$u'(c_0) = (1+r)u'(c_1) + \mu \tag{1.40}$$

that is, with a binding liquidity constraint ($\mu > 0$), the marginal utility of consumption decreases over time. This is because credit rationing forces the individual to consume *less* in the first period and *more* in the second.

Note that if $\mu = 0$, the liquidity constraint is *slack*, and the Euler equation seems to be the same as in the case of perfect financial markets. A *crucial implication* of the model with liquidity constraints is, therefore, that the rate of growth of consumption is, on average, higher in this case than with no constraints. We will return to this aspect in the subsequent chapters.

1.5 T-Period Generalization with Discrete Time

Now, we return to the hypothesis of perfect capital markets and generalize the two-period model in Sect. 1.2. The simplest way is to rewrite the first-order conditions for the general case of optimization between t and $t+1$:

$$u'(c_t) = \lambda\left(\frac{1+\delta}{1+r}\right)^t \tag{1.41}$$

and

$$u'(c_{t+1}) = \lambda\left(\frac{1+\delta}{1+r}\right)^{t+1}, \tag{1.42}$$

where, with $u'(c_t)$, we indicate the first derivative with respect to c_t, and with $u'(c_{t+1})$, the corresponding derivative with respect to c_{t+1}.

Taking logs of Eqs. 1.41 and 1.42 we get

$$\ln u'(c_t) = \ln \lambda + t \ln \left(\frac{1+\delta}{1+r} \right) \tag{1.43}$$

and

$$\ln u'(c_{t+1}) = \ln \lambda + (t+1) \ln \left(\frac{1+\delta}{1+r} \right). \tag{1.44}$$

Subtracting Eq. 1.43 and Eq. 1.44 yields

$$\Delta \ln u'(c_{t+1}) = \ln \left(\frac{1+\delta}{1+r} \right). \tag{1.45}$$

This last equation can be further simplified through a Taylor expansion of the term $\ln u'(c_{t+1})$ and approximating by $\ln(1+x) \simeq x$, under the assumption that $x \ll 1$.

Expanding the term $\ln u'(c_{t+1})$ to the first order in the neighborhood of c_t and multiplying and dividing by c_t gives

$$c_t \frac{u''(c_t)}{u'(c_t)} \frac{c_{t+1} - c_t}{c_t} \simeq (\delta - r). \tag{1.46}$$

From this, it follows that

$$\frac{\Delta c}{c} \simeq \sigma (r - \delta), \tag{1.47}$$

where $\sigma = -u'(c_t)/c_t u''(c_t)$ measures the *elasticity of intertemporal substitution* (EIS). Note that since, by assumption, $u'(c_t) > 0$ and $u''(c_t) < 0$, then $\sigma > 0$. Note also that, as before, the dynamics of consumption depends on the relationship between δ and r; indeed, we have that $\Delta c \lesseqqgtr 0$ when $\delta \gtreqqless r$.

1.5.1 EIS and Risk Aversion

The form of the utility function affects the relationship that links *risk aversion*[2] with EIS. Suppose, for example, that the utility function exhibits *constant relative risk aversion* (CRRA):

$$u(c_t) = \frac{c_t^{1-\gamma}}{1-\gamma}, \tag{1.48}$$

[2] See Chap. 2 for a more in-depth discussion about risk aversion.

with $\gamma > 0$ and with $u^{'}(c_t) = c_t^{-\gamma}$. The *degree* of relative risk aversion is measured by the *index* of relative risk aversion:

$$-c_t \frac{u^{''}(c_t)}{u^{'}(c_t)} = \gamma, \tag{1.49}$$

which is equal to the constant γ. The Euler equation for the function in Eq. 1.48 thus becomes

$$\left(\frac{c_{t+1}}{c_t}\right)^{\gamma} = \left(\frac{1+r}{1+\delta}\right); \tag{1.50}$$

that is, approximating by logarithms,

$$\Delta \ln c_{t+1} = \frac{1}{\gamma}(r - \delta). \tag{1.51}$$

It thus follows that for a CRRA utility function, the elasticity of intertemporal substitution is equal to σ, which is the reciprocal of relative risk aversion ($\sigma = 1/\gamma$).

1.6 A Formal Model

The solution for T periods in discrete time can also be derived using the Lagrangian function, as in the two-period model. We start with the following problem in general form:

$$\begin{cases} \max\limits_{c_t} \sum_{s=0}^{T-t} \frac{u(c_{t+s})}{(1+\delta)^s} \\ \text{s.t.} \quad a_{t+s+1} = (1+r)(a_{t+s} + y_{t+s} - c_{t+s}) \\ \qquad (1+r)^{-T} a_{T+1} = 0. \end{cases} \tag{1.52}$$

The constraint in Eq. 1.52 requires that the present value of wealth at time $T+1$ is zero, which, in turn, implies that $a_{T+1} = 0$, being $(1+r)^{-T} > 0$. As before, this is the *transversality condition*.

We explore the economic meaning of this problem by positing that $s = 0$. We rewrite the constraint Eq. 1.52 as

$$a_t = \frac{1}{1+r} a_{t+1} - y_t + c_t. \tag{1.53}$$

This is a first-order difference equation in a_t that can be solved by iteratively substituting forward for a_{t+s} until time T. In detail, when $s = 1$ we have

$$a_{t+1} = \frac{1}{1+r} a_{t+2} - y_{t+1} + c_{t+1}. \tag{1.54}$$

By substituting a_{t+1} into Eq. 1.53, we obtain

$$a_t = \frac{1}{1+r}\left(\frac{1}{1+r} a_{t+2} - y_{t+1} + c_{t+1}\right) - y_t + c_t; \tag{1.55}$$

that is, simplifying

$$a_t = \frac{1}{(1+r)^2} a_{t+2} + \frac{c_{t+1}}{1+r} + c_t - \frac{y_{t+1}}{1+r} - y_t. \tag{1.56}$$

By iterating until $s = T - t$, we get

$$\sum_{s=0}^{T-t} \frac{c_{t+s}}{(1+r)^s} + \frac{a_{T+1}}{(1+r)^T} = a_t + \sum_{s=0}^{T-t} \frac{y_{t+s}}{(1+r)^s}. \tag{1.57}$$

However, the transversality condition $(1+r)^{-T} a_{T+1} = 0$ implies that, over the planning horizon, the discounted value of consumption must equal the discounted value of future income, plus the initial value of wealth a_t (if any).

Note that, as in the two-period model, if the transversality condition were positive, that is, if $(1+r)^{-T} a_{T+1} > 0$, the present value of consumption would be *smaller* than the discounted value of income since

$$\sum_{s=0}^{T-t} \frac{c_{t+s}}{(1+r)^s} = a_t + \sum_{s=0}^{T-t} \frac{y_{t+s}}{(1+r)^s} - \frac{a_{T+1}}{(1+r)^T}. \tag{1.58}$$

But, the optimization criterion states that consumer welfare is maximized when consumption has a present value equal to that of total resources. Thus, it would be *suboptimal* to maintain positive wealth $a_{T+1} > 0$ at the end of the planning horizon. This implies that $\frac{a_{T+1}}{(1+r)^T} = 0$.

Analogously, and again in keeping with the two-period model, it is not possible to have a negative level of wealth in the last period, $a_{T+1}/(1+r)^T < 0$. This would imply that the present value of consumption is *higher* than that of future income. In other words, the consumer will continuously go into debt to repay previous debts (debt rollover) rather than repaying his or her creditors. As a result, the debt would grow at a rate equal to the interest factor $(1+r)$.

It is thus *necessary* that at the end of the planning horizon, the wealth is zero; that is, $a_{T+1} = 0$. This condition is known in the literature as the *no-Ponzi-game condition*.

The Lagrangian of Eq. 1.52 can thus be written as

$$L_t = \sum_{s=0}^{T-t} \frac{u(c_{t+s})}{(1+\delta)^s} - \lambda \left(\sum_{s=0}^{T-t} \frac{c_{t+s}}{(1+r)^s} - a_t - \sum_{s=0}^{T-t} \frac{y_{t+s}}{(1+r)^s} \right), \tag{1.59}$$

where λ is the Lagrange multiplier. By differentiating L_t in t and $t+s$ in Eq. 1.59, we obtain

$$u'(c_t) = \lambda. \tag{1.60}$$

$$\frac{u'(c_{t+s})}{(1+\delta)^s} = \frac{\lambda}{(1+r)^s}; \tag{1.61}$$

that is

$$u'(c_t) = \frac{(1+r)^s}{(1+\delta)^s} u'(c_{t+s}). \tag{1.62}$$

If $u(c_t) = \frac{c_t^{1-\gamma}}{1-\gamma}$ and $s = 1$, the Euler equation becomes

$$\left(\frac{c_{t+1}}{c_t}\right)^\gamma = \left(\frac{1+r}{1+\delta}\right), \tag{1.63}$$

which is equal to the Euler equation derived in Sect. 1.2.

1.6.1 On the Transversality Condition

We further clarify the economic meaning of the transversality condition in Eq. 1.52 by focusing on the Euler condition, Eq. 1.62. We solve this first-order difference equation by the method of substitutions up to time T.

Starting from $t+1$, the first forward substitution implies that

$$u'(c_t) = \frac{1+r}{1+\delta}\left[\frac{1+r}{1+\delta} u'(c_{t+2})\right] = \left(\frac{1+r}{1+\delta}\right)^2 u'(c_{t+2}). \tag{1.64}$$

Then, substituting iteratively until T, we get

$$u'(c_t) = \left(\frac{1+r}{1+\delta}\right)^T u'(c_T) \tag{1.65}$$

This condition is consistent with Eq. 1.52. In fact, if no transversality condition is imposed on the budget constraint, then

$$\lambda\left(\sum_{s=0}^{T-t} \frac{c_{t+s}}{(1+r)^s} - a_t - \sum_{s=0}^{T-t} \frac{y_{t+s}}{(1+r)^s}\right) = 0; \tag{1.66}$$

that is,

$$\lambda\left(\frac{a_{T+1}}{(1+r)^T}\right) = 0 \tag{1.67}$$

Now, using the first-order condition $\lambda = u'(c_t)$, this expression becomes

$$u'(c_t)\left(\frac{a_{T+1}}{(1+r)^T}\right) = 0. \tag{1.68}$$

Since we have assumed that $u'(c_t)$ is strictly positive, this implies that a_{T+1} must be zero to satisfy the optimality condition, Eq. 1.65.

Thus, transversality is a *necessary* condition for solving the dynamic optimization problem (Kamihigashi 2002). It can take several forms:

$$(1+r)^{-T} a_{T+1} = 0, \tag{1.69}$$

$$u'(c_t)(1+r)^{-T} a_{T+1} = 0, \tag{1.70}$$

$$\lambda a_{T+1} = 0. \tag{1.71}$$

1.7 Consumption in Continuous Time

In the models presented thus far, we have assumed that time passes at *discrete* intervals. We now extend the analysis to the case of *continuous time*. Consider the case of a consumer with the following CRRA function:

$$\begin{cases} \max. & \int_0^T \frac{c_t^{1-\gamma}}{1-\gamma} \exp(-\delta t)\, dt \\ \text{s.t.} & \dot{a}_t = r a_t + y_t - c_t \\ & (1+r)^{-T} a_T = 0, \end{cases} \tag{1.72}$$

where $c_t > 0$ and where $\dot{a}_t = da_t/dt$. To solve Eq. 1.72 in continuous time, it is useful to apply the *maximum principle* , which is a Lagrangian-equivalent solution (see Appendix 1 at the end of the chapter) known as the Hamiltonian function. Formally,

$$H_t = \frac{c_t^{1-\gamma}}{1-\gamma} \exp(-\delta t) + \lambda_t (r a_t + y_t - c_t), \tag{1.73}$$

where λ_t is referred to as the *costate* variable and is a measure of the shadow value of the constraint. The first-order conditions are the following:

$$\frac{\partial H}{\partial c} = 0 : \; c_t^{-\gamma} \exp(-\delta t) = \lambda_t. \tag{1.74}$$

$$-\frac{\partial H}{\partial a} = \dot{\lambda} : \; r = \frac{\dot{\lambda}}{\lambda_t}. \tag{1.75}$$

$$\frac{\partial H}{\partial \lambda} = \dot{a} : \; \dot{a}_t = r a_t + y_t - c_t \tag{1.76}$$

The transversality condition is

$$\lambda_T a_T = 0 \tag{1.77}$$

From Eq. 1.74, at time T

$$c_T^{-\gamma} \exp(-\delta T) = \lambda_T > 0. \tag{1.78}$$

Since consumption is positive at each instant t, we impose the requirement that at the end of the planning horizon $a_T = 0$ in order to satisfy the condition in Eq. 1.77.

The growth path of consumption can be derived from Eq. 1.75 by substituting the expressions of λ_t and $\dot{\lambda}$ from Eq. 1.74. We get

$$r = -\frac{-\delta c_t^{-\gamma} \exp(-\delta t) - \gamma c_t^{-(\gamma+1)} \exp(-\delta t)\, \dot{c}}{c_t^{-\gamma} \exp(-\delta t)}; \tag{1.79}$$

that is,

$$\frac{\dot{c}}{c_t} = \frac{1}{\gamma}(r - \delta). \tag{1.80}$$

Equation 1.80 is a first-order differential equation with constant coefficients that we can rewrite as

$$c_t = \left(\frac{\gamma}{r - \delta}\right) \dot{c} \tag{1.81}$$

the solution is

$$c_t = c_0 \exp\left[\left(\frac{r - \delta}{\gamma}\right) t\right]. \tag{1.82}$$

Thus, as before, for a given value of γ (i.e., for a given EIS), the dynamics of consumption may increase or decrease depending on whether r is larger or smaller than δ. This is an important result, but it leaves still the following question open: although Eq. 1.82 describes the dynamics of c_t, it does not provide information about the economic variables affecting consumption over time. We consider this issue in the next section.

1.7.1 The Consumption Function

We answer the question above by integrating the budget constraint $\dot{a}_t = ra_t + y_t - c_t$ with respect to time, after multiplying it by $\exp(-rt)$. We obtain

$$\int_0^T (\dot{a}_t - ra_t) \exp(-rt) dt = \int_0^T y_t \exp(-rt) dt - \int_0^T c_t \exp(-rt) dt. \tag{1.83}$$

Given that the primitive of $(\dot{a}_t - ra_t) \exp(-rt)$ is the exponential function $a_t \exp(-rt)$, it follows that

$$(a_t \exp(-rt))_0^T = a_T \exp(-rT) - a_0 = \int_0^T y_t \exp(-rt) dt - \int_0^T c_t \exp(-rt) dt; \tag{1.84}$$

that is,

$$\int_0^T c_t \exp(-rt) dt = a_0 + \int_0^T y_t \exp(-rt) dt \equiv a_0 + h_0, \tag{1.85}$$

where h_0 denotes the discounted value of future income, and where $a_T = 0$ to satisfy the transversality condition.

We now substitute Eq. 1.82 into Eq. 1.85 to get the consumption function:

$$\int_0^T c_0 \exp(-\alpha t)dt = a_0 + h_0 \tag{1.86}$$

where, to simplify the notation, we put $\alpha = \left(r - \frac{r-\delta}{\gamma}\right) > 0$.

Finally, solving the integral in Eq. 1.86 we have

$$\left[-\frac{c_0 \exp(-\alpha t)}{\alpha}\right]_0^T = a_0 + h_0. \tag{1.87}$$

that is,

$$-\frac{c_0 \exp(-\alpha T)}{\alpha} + \frac{c_0}{\alpha} = a_0 + h_0, \tag{1.88}$$

from which we get the *consumption function*

$$c_0 = \frac{\alpha}{1 - \exp(-\alpha T)} (a_0 + h_0) = \beta (a_0 + h_0). \tag{1.89}$$

The parameter $\beta = \frac{\left(r - \frac{r-\delta}{\gamma}\right)}{1-\exp(-\alpha T)} < 1$ can be understood as the *marginal propensity to consume*. It depends on parameters, time, but not on wealth. However, consumption is a linear function of wealth, given by the sum of initial wealth a_0 and the present value h_0 of future income. Therefore, the distribution of resources over time *does not affect* consumption choices.

Notably, Eq. 1.89 is different from that postulated by Keynes (1936) in his General Theory, according to which consumption is related to current income and not to wealth and future income. Instead, the consumption equation derived here is very much like that proposed by Friedman (1957), which traces Keynesian consumption to the neoclassical framework. Traditionally, in consumption theory, this formulation is known as the *permanent income hypothesis*, according to which consumption varies "smoothly" over time as income and wealth change. This is because individuals desire a consistent standard of living and will gradually adjust their spending patterns in response to changes in their financial circumstances.

Next, we analyze what happens when $T \to \infty$. In this case the consumption function becomes $c_0 = \alpha (a_0 + h_0)$. If, in addition, $\delta = r$ the marginal propensity to consume is equal to the interest rate r:

$$c_0 = r (a_0 + h_0). \tag{1.90}$$

That is, consumption is constant over time and equal to the annuity of wealth. In this formulation, r captures the perpetual income factor of wealth. As a final remark, note that the effect of an interest rate change on consumption is ambiguous. If we consider the case of $T \to \infty$ and calculate the derivative of α with respect to r we have

$$\frac{\partial \alpha}{\partial r} = \left[1 - \frac{1}{\gamma}\right]. \tag{1.91}$$

It is easy to show that the algebraic sign of Eq. 1.91 depends on the reciprocal of the relative risk aversion $1/\gamma$. If, for example, $\gamma = 1$, a change in the interest rate leaves the marginal propensity to consume unchanged. This is the case of logarithmic utility in which the income and substitution effects exactly *offset* each other. Alternatively, if $\gamma < 1$, the income effect prevails, and the derivative is positive; that is, current consumption increases and wealth decreases.

Note that when $T \to \infty$, the transversality condition takes the form:

$$\lim_{t\to\infty} a_t \, exp\,(-rt) = 0. \tag{1.92}$$

1.7.2 The Saving Function

The previous result implies that consumption in one period does not depend on current income but on the current value of income received by an individual over his or her entire *life horizon*. As stated, the expression $\beta\,(a_0 + h_0)$ is consistent with the definition of *permanent income*, and the difference between this value and current income measures *temporary income*.

As noted above, Friedman's predictions of consumption smoothing—that is, individuals prefer to distribute transitory changes in income over time—depart from the traditional Keynesian emphasis on a higher marginal propensity to consume from current income. We can appreciate the importance of this distinction by considering the case where, at time t, income transitorily increases by an amount Δy. Although current income is then $y_t + \Delta y$, consumption only increases by a share equal to $\beta \Delta y = \left[\frac{\alpha}{1-\exp(-\alpha T)}\right] \Delta y$. Therefore, the longer the consumer's planning horizon (i.e., the more distant time T), the lower the propensity to consume β and the smaller the impact of the changes in current income on current consumption.

The second implication of our result above concerns the *saving function*. In fact, although current income is not important in determining current consumption, it has an important influence on saving. By definition, savings s_t is the difference between disposable and consumed income

$$s_t = y_t - c_0. \tag{1.93}$$

Substituting Eq. 1.90 in Eq. 1.93, we obtain

$$s_t = (y_t - \beta h_0) - \beta a_0. \tag{1.94}$$

The latter expression states that savings increase when current income y_t is relatively higher than average income βh_0, that is when positive transitory shocks cause a short-term increase in the individual's level of income. Conversely, saving decreases when current income falls below average income. The main consequence of this behavior is that the consumer uses his or her saving as a *buffer stock* to keep consumption essentially *stable* over time.

In conclusion, since the consumption function, Eq. 1.89, is the result of the intertemporal optimal choice function derived from the Euler equation, saving acts as a balancing factor, ensuring a consumption path characterized by an absence of jumps and discontinuities. If this were not the case, this would imply that between successive periods, the condition of equality between the MRS and prices is violated, signaling the suboptimality of the consumption choice. This behavior is known in the literature as *consumption smoothing*.

Appendix 1

Hamiltonian Function and Optimal Control

In this appendix, we present the essential elements of deterministic optimal control theory and the Hamiltonian function. The Maximum Principle is the central element of this methodology. Let us consider a dynamic system. We divide the variables into two categories:

- *state* variables, for example, the wealth of individuals (or the capital stock of firms);
- *control* variables, for example, consumption (or investment).

There is a well-determined cause-and-effect relationship between these variable types: once the value of the control variable is chosen, the state variable in each period is determined accordingly. For example, by choosing its optimal (control) consumption period by period, a consumer determines the dynamics of its wealth stock, given the initial endowment and the planning interval.

Obviously, the key issue is to choose an optimal control sequence that ensures maximization (or minimization) of the objective function (e.g., intertemporal utility).

The purpose of optimal control is to derive a sequence of optimal values for the control variable at each instant. In other words, the principle requires finding a functional form that, over the entire optimization interval (program), describes the optimal evolution of the control variable (and thus of the corresponding state variable) and guarantees the maximum (minimum) value of the objective function.

In its simplest form, an optimal control problem can be written as follows. For each t find the level of consumption $c(t)$ that maximizes: [3]

$$\begin{cases} \text{max.} & \int_0^T v\left(c(t), s(t)\right) dt \\ \text{s.t.} & \dot{s} = g\left[s(t), c(t)\right] \\ & s(0) = s_0 \\ & s(T) = s_T, \end{cases} \tag{A.1}$$

where $s(t)$ is the state variable, $\dot{s} = \frac{ds}{dt}$ is its variation with respect to time, $c(t)$ is the control variable, and $[0, T]$ is the programming horizon. We denote by s_0 and s_T the initial and final values of the state variable. If the control $c(t)$ is optimally chosen over the entire interval $[0, T]$, the differential equation for $\dot{s}$ and conditions for $s(0)$ and $s(T)$ are sufficient to determine the evolution of $s(t)$ over the programming horizon. In fact, the (maximum) value of $V\left[c(t), s(t)\right]$ depends on the latter variable.

The optimal control is assumed to exist and be unique and differentiable with respect to time. The necessary conditions that constitute the Maximum Principle require that we introduce a convenience variable, similar to a Lagrange multiplier, known as the costate variable, denoted by $\lambda(t)$. Therefore, at each instant, we define a new function—the Hamiltonian function—which, for our problem, is written as follows:

$$H\left[c(t), s(t), \lambda(t)\right] = v\left[c(t), s(t)\right] + \lambda(t) g\left[s(t), c(t)\right]. \tag{A.2}$$

Maximum Principle

An optimal solution to the problem above is, for each instant t, a set of *three values* $c(t), s(t), \lambda(t)$ that satisfies the following conditions:

1. $c(t)$ maximizes $H\left[c(t), s(t), \lambda(t)\right]$; that is,

$$\frac{\partial H}{\partial c(t)} = 0. \tag{A.3}$$

2. The costate variable $\lambda(t)$ and state variable $s(t)$ satisfy the following pair of differential equations:

[3] Note that in this Appendix, in that of Chap. 4 and in Sect. 5.5 of the book we use the notation $x(t)$ instead of x_t for clarity of exposition.

$$\dot{\lambda}(t) = -\frac{\partial H}{\partial s(t)}, \tag{A.4}$$

$$\dot{s}(t) = \frac{\partial H}{\partial \lambda(t)}, \tag{A.5}$$

with the two boundary conditions in Eq. A.1. Therefore, the optimal conditions consist of two differential equations describing the dynamics of the costate variable $\lambda(t)$ and the state $s(t)$, plus an algebraic equation obtained from the first-order condition that allows for the optimal choice of control $c(t)$.

Further Reading

The topics dealt with in this chapter are discussed in many textbooks and papers dedicated to consumption theory. These include the two classic Varian (1992), Varian (2014) textbooks: *Microeconomic analysis* and *Intermediate Microeconomics*. Particularly rich in content is Deaton (1992) and the survey on household savings theories by Browning and Lusardi (1996). For the macroeconomic implications of consumption decisions, the reader may instead refer to Blanchard and Fischer (1989) (Chaps. 1 and 6) and Romer (2019) (Chap. 8). Very interesting for its empirical evidence is the book by Jappelli and Pistaferri (2017). A good review of the theory of optimal control is offered in Acemoglu (2009) (Chaps. 6 and 7), and a number of different applications are given in Stokey and Lucas (1989) (Chaps. 4 and 5). For a comprehensive introduction to dynamic models with difference and differential equations, see Lambert (1985), Leonard and Van Long (1992), Chiang and Wainwright (2005), and Gandolfo (2010). For the debate on the form of the consumption function, see Keynes (1936), Friedman (1957) and Guilfoil (1962). Finally, for a review of empirical studies on precautionary savings and the problems associated with measuring risk and wealth see Kennickell and Lusardi (2004).

References

Acemoglu, D. (2009). Introduction to modern economic growth. Prince- ton University Press.

Blanchard, O. J., & Fischer, S. (1989). Lectures on macroeconomics. MIT University Press.

Browning, M., & Lusardi, A. (1996). Household saving: Micro theories and micro facts. Journal of Economic literature, 34 (4), 1797- 1855.

Calcagnini, G., & Saltari, E. (2009). The economics of imperfect mar- kets. the effects of market imperfections on economic decision- making.

Chiang, A. C., & Wainwright, K. (2005). Fundamental methods of math- ematical economics. Erlangga.

Deaton, A. (1992). Understanding consumption. Oxford University Press.

Fisher, I. (1930). The theory of interest as determined by impatience to spend income and opportunity to invest it. The Macmillan Company.

Friedman, M. (1957). Theory of the consumption function. Princeton University Press.

Gandolfo, G. (2010). Economic dynamics. Springer.

Guilfoil, J. D. (1962). A survey of consumption theory-from keynes through friedman. The American Economist, 6 (1), 12-20.

Hurst, E., & Lusardi, A. (2004). Liquidity constraints, household wealth, and entrepreneurship. Journal of political Economy, 112 (2), 319- 347.

Jappelli, T., & Pistaferri, L. (2017). The economics of consumption: Theory and evidence. Oxford University Press.
Kamihigashi, T. (2002). A simple proof of the necessity of the transver- sality condition. Economic theory, 20, 427-433.
Karush, W. (1939). Minima of functions of several variables with in- equalities as side constraints. M. Sc. Dissertation. Dept. of Math- ematics, Univ. of Chicago.
Kennickell, A., & Lusardi, A. (2004). Disentangling the importance of the precautionary saving mode.
Keynes, J. M. (1936). The general theory of employment, interest and money. Palgrave Macmillan.
Kuhn, H. W., & Tucker, A. W. (1951). Nonlinear programming. Pro- ceedings, Second Berkeley Symposium in Mathematical Statis- tics; Probability, J. Neyman, editor. University of California Pres, Berkeley.
Lambert, P. J. (1985). Advanced mathematics for economists: Static and dynamic optimization. B. Blackwell.
Leonard, D., & Van Long, N. (1992). Optimal control theory and static optimization in economics. Cambridge University Press.
Ramsey, F. P. (1928). A mathematical theory of saving. The Economic Journal, 38 (152), 543-559.
Reuben, E., Sapienza, P., & Zingales, L. (2010). Time discounting for primary and monetary rewards. Economics Letters, 106 (2), 125- 127.
Romer, D. (2019). Advanced macroeconomics, 5th edition. McGraw-Hill Esducation.
Stokey, N. L., & Lucas, R. E. (1989). Recursive methods in economic dynamics. Harvard University Press.
Varian, H. R. (1992). Microeconomic analysis. third edition. WW norton & company.
Varian, H. R. (2014). Intermediate microeconomics: A modern approach, ninth edition. WW norton & company.
Zeldes, S. P. (1989). Optimal consumption with stochastic income: De- viations from certainty equivalence. The Quarterly Journal of Economics, 104 (2), 275-298.

2 Consumption and Uncertainty in Discrete Time

- In this chapter, we introduce two forms of *uncertainty*: uncertainty about *labor incomes* and uncertainty about *asset returns*.
- The intertemporal problem is solved using *dynamic programming*.
- When labor income is uncertain, we obtain the well-known result that the level of consumption follows a *random walk*.
- When returns are uncertain, we show that the allocation of savings among different risky assets depends on the *utility function* and the *probability distribution of returns*.
- We then introduce the concept of *risk aversion* and discuss *precautionary saving*.
- Finally, we derive a relationship linking *intertemporal consumption* and *portfolio choice*.

2.1 Introduction

Uncertainty can take many forms, relating to diverse factors, including regulatory changes, geopolitical events, consumer preferences, and/or environmental factors. Each of these can have a significant impact on consumers' decision-making processes. We set up the discussion by focusing on three specific types of uncertainty concerning the following:

1. *Future income* (the consumer is uncertain about the level of future labor income);
2. *Return on assets* (the consumer is uncertain about the rate of return on assets that make up his or her wealth);
3. *Both apply*.

G. Travaglini et al., *Notes on Consumption Theory*, Classroom Companion: Economics,
https://doi.org/10.1007/978-3-031-54986-1_2

We start with the first scenario; that is, we assume that future income is uncertain, and the consumer must make a probabilistic assessment of its expected value. This hypothesis, together with the simplifying assumption of *linear marginal utility* of consumption, allows us to derive a particular form of intertemporal consumption under uncertainty known as the certainty equivalence model. As we see below, although historically important, this formalization of the consumption function has limited empirical application; it only allows us to analyze some of the effects of income uncertainty on consumption and saving behaviors. We now consider this in detail.

2.2 Uncertain Income

In accordance with the model developed in the previous chapter, we assume that capital markets are perfect. Only one riskless asset exists. It provides a return equal to r. Time flows discretely, and the planning horizon is finite. We assume, however, that there is uncertainty about the level of future labor income, y_{t+s}. In short, future labor income is a *stochastic* variable.

The *expected* utility maximization problem can then be written as

$$\begin{cases} \max. & E_t\left[\sum_{s=0}^{T-t} \frac{u(c_{t+s})}{(1+\delta)^s}\right] \\ \text{s.t.} & a_{t+s+1} = (1+r)\left(a_{t+s} + y_{t+s} - c_{t+s}\right), \end{cases} \tag{2.1}$$

where $u\left(c_{t+s}\right)$ is an additive and separable utility function and $E_t\left[\cdot\right]$ denotes an individual's expectations about his or her future income, conditional on the information available at time t. As in the problems considered above, we assume that the initial wealth a_t is given and that the following transversality condition holds:

$$(1+r)^{-T} a_{T+1} = 0. \tag{2.2}$$

2.2.1 Dynamic Programming

Although the Lagrangian function can be applied, we solve Eq. 2.1 using dynamic programming. Therefore, we define the value function

$$V_t\left(a_t\right) = \max_{c_t} E_t\left[\sum_{s=0}^{T-t} \frac{u\left(c_{t+s}\right)}{(1+\delta)^s}\right], \tag{2.3}$$

where $V_t\left(a_t\right)$ represents the present value of expected utility that the consumer obtains from optimal consumption, given the initial wealth a_t. Consumption c_{t+s} is the *control* variable, and a_{t+s} is the *state* variable, the evolution of which depends on consumption choices.

With the definition of $V_t\left(a_t\right)$, the multiperiodic problem reduces to a sequence of recursive biperiodic problems that can be expressed by the following *Hamilton-Jacobi-Bellman equation* (HJB) (Bellman 1957)

$$V_t\left(a_t\right)=\max_{c_t}\left\{u\left(c_t\right)+\frac{1}{1+\delta}E_t\left[V_{t+1}\left(a_{t+1}\right)\right]\right\}. \tag{2.4}$$

In this expression, $u\left(c_t\right)$ is the utility obtained at the present time from consuming c_t, while $\frac{1}{1+\delta}E_t\left[V_{t+1}\left(a_{t+1}\right)\right]$ is the present value of the expected utility obtained by the consumer in the remainder of the period by choosing an appropriate level of future consumption.

For purposes of clarity, suppose that the planning horizon begins in t and ends in $t+1$. Let us consider the maximization problem in the final period $t+1$. The goal is to maximize $u\left(c_{t+1}\right)$ by appropriately choosing c_{t+1}, given the constraint. The intertemporal separability of the utility function and the recursive structure of the budget constraint allows us to derive a solution that provides the optimal value of $c^*_{t+1}\left(a_{t+1}\right)$.

This level of consumption maximizes the objective function in $t+1$; that is, it maximizes the value of $V_{t+1}\left(a_{t+1}\right)$. Once the solution is obtained, we proceed backward by solving for the optimum consumption in t using the same procedure, knowing the optimal level of wealth in $t+1$.

By substituting the constraint a_{t+1} from Eq. 2.1 into Eq. 2.4 we have

$$V_t\left(a_t\right)=\max_{c_t}\left\{u\left(c_t\right)+\frac{1}{1+\delta}E_t\left[V_{t+1}\left(1+r\right)\left(a_t+y_t-c_t\right)\right]\right\}. \tag{2.5}$$

The first-order condition with respect to c_t is

$$u'\left(c_t\right)=\frac{1+r}{1+\delta}E_t\left[V'_{t+1}\left(a_{t+1}\right)\right]. \tag{2.6}$$

Unfortunately, it is impossible to directly solve the condition expressed in Eq. 2.6, which in this context corresponds to the Euler equation, because the form of the value function $V_t\left(a_t\right)$ is unknown. However, there is an envelope relationship between $V'_t\left(a_t\right)$ and $u'\left(c_t\right)$ along the optimal consumption path that can be used to solve the first-order condition.

We thus consider what happens to Eq. 2.5 when individual wealth a_t varies infinitesimally. Keep in mind that the value of $V_t\left(a_t\right)$ expresses the value of utility obtained after solving the optimization problem. Equation 2.5 is the *indirect utility function*, which is obtained by substituting the sequence of optimal solutions $c^*_{t+s}\left(a_{t+s}\right)$ into the expected utility function $u(c_t)$.

Formally, differentiating with respect to a_t yields

$$V'_t\left(a_t\right)=u'\left(c_t\right)\frac{\partial c_t}{\partial a_t}+\frac{1+r}{1+\delta}E_t\left[V'_{t+1}\left(a_{t+1}\right)\right]-\frac{1+r}{1+\delta}E_t\left[V'_{t+1}\left(a_{t+1}\right)\right]\frac{\partial c_t}{\partial a_t}. \tag{2.7}$$

That is,

$$V'_t\left(a_t\right)=\left[u'\left(c_t\right)-\frac{1+r}{1+\delta}E_t\left(V'_{t+1}\left(a_{t+1}\right)\right)\right]\frac{\partial c_t}{\partial a_t}+\frac{1+r}{1+\delta}E_t\left[V'_{t+1}\left(a_{t+1}\right)\right]. \tag{2.8}$$

However, from the first-order condition, it follows that

$$u'(c_t) - \frac{1+r}{1+\delta} E_t \left[V'_{t+1}(a_{t+1}) \right] = 0. \tag{2.9}$$

Therefore, the envelope with respect to a_t implies

$$V'_t(a_t) = \frac{1+r}{1+\delta} E_t \left[V'_{t+1}(a_{t+1}) \right]. \tag{2.10}$$

The latter expression solves Eq. 2.6. In fact, by substituting this expression into the Euler equation, we get

$$u'(c_t) = V'_t(a_t); \tag{2.11}$$

that is, along the optimal trajectory, the marginal utility of consumption coincides with the marginal utility of wealth. Clearly, this optimality relation must be valid in each period and thus also in $t+1$:

$$u'(c_{t+1}) = V'_t(a_{t+1}). \tag{2.12}$$

By substituting Eq. 2.12 into Eq. 2.6 we get

$$u'(c_t) = \frac{1+r}{1+\delta} E_t u'(c_{t+1}). \tag{2.13}$$

This is the Euler equation between two successive time periods: in equilibrium, an increase in marginal utility caused by a reduction in consumption in t is offset by a reduction in marginal utility in $t+1$ caused by an increase in c_{t+1}. Finally, it is useful to use a special form of Eq. 2.13 that occurs when $\delta = r$. In this case, the condition becomes

$$u'(c_t) = E_t u'(c_{t+1}). \tag{2.14}$$

This final equation states that, at the present time, the expected value of marginal utility in $t+1$ is equal to the present value of marginal utility. Thus, the *best predictor* of marginal utility in $t+1$ is $u'(c_t)$ from the previous period.

2.2.2 The Case of Quadratic Utility

Using Eq. 2.13, the Euler equation can be rewritten as

$$\frac{1+\delta}{1+r} u'(c_t) = E_t u'(c_{t+1}) \tag{2.15}$$

Now, considering a quadratic utility function of the form $u(c_t) = ac_t - \frac{b}{2}c_t^2$, Eq. 2.15 becomes

$$\frac{1+\delta}{1+r}(a - bc_t) = E_t(a - bc_{t+1}), \tag{2.16}$$

where, to avoid negative marginal utility, the condition $0 < c_t < a/b$ must be imposed. Rearranging, we get

$$E_t\left(c_{t+1}\right) = \alpha + \beta c_t, \tag{2.17}$$

with $\alpha = \frac{a}{b}\frac{r-\delta}{1+r}$ and $\beta = \frac{1+\delta}{1+r}$; that is, consumption follows a first-order autoregressive process AR(1). Further, notice that if expectations are rational and the shock ε_t on consumption follows the normal distribution $\varepsilon_t \backsim N\left(0, \sigma^2\right)$, we can write that ex ante:

$$c_{t+1} = \alpha + \beta c_t + \varepsilon_{t+1} \tag{2.18}$$

since, by assumption, $E_t\left(\varepsilon_t\right) = 0$. Finally, if $\delta = r$

$$c_{t+1} = c_t + \varepsilon_{t+1}; \tag{2.19}$$

that is,

$$E_t\left(c_{t+1}\right) = c_t. \tag{2.20}$$

The main implication of this result is that consumption is a *random walk* and the best prediction of tomorrow's consumption is today's consumption (Hall 1978).

2.2.3 The Certainty Equivalence Model

Since y_t is a stochastic variable, we must now compute the *expected* value of the budget constraint[1]

$$\sum_{s=0}^{T-t} \frac{E_t\left(c_{t+s}\right)}{(1+r)^s} = a_t + \sum_{s=0}^{T-t} \frac{E_t\left(y_{t+s}\right)}{(1+r)^s} \tag{2.21}$$

In the case of quadratic utility, from the first-order condition, we know that if $\delta = r$, then $E_t\left(c_{t+1}\right) = c_t$. Thus, consumption remains constant over time; that is, $c_t = c_{t+1} = \ldots = c_T$.

Substituting this latter condition into Eq. 2.21 we have

$$c_t \sum_{s=0}^{T-t} \frac{1}{(1+r)^s} = a_t + \sum_{s=0}^{T-t} \frac{E_t\left(y_{t+s}\right)}{(1+r)^s} \tag{2.22}$$

Since

$$\sum_{s=0}^{T-t} \frac{1}{(1+r)^s} = \left[1 - \frac{1}{(1+r)^{T-t+1}}\right] \frac{1+r}{r} \equiv \frac{1}{\theta} \tag{2.23}$$

[1] Again, the solution for the dynamics of the budget constraint is obtained by recursively substituting up to T for the value of wealth. For the model in the case of certainty, see Sect. 1.3 of Chap. 1.

consumption c_t can be rewritten as

$$c_t = \theta \left[a_t + \sum_{s=0}^{T-t} \frac{E_t\left(y_{t+s}\right)}{(1+r)^s} \right] = \theta \left(a_t + E_t\left(h_t\right)\right). \tag{2.24}$$

This is the solution of Eq. 1.89, but under the assumption of uncertain labor income; as before θ measures the marginal propensity to consume. Moreover, when $T \to \infty$

$$c_t = \frac{r}{1+r}\left[a_t + E_t\left(h_t\right)\right], \tag{2.25}$$

where the term to the right-hand side is the expected permanent income. Equation 2.25 is the fundamental equation of the *Certainty Equivalence model*. The ratio $\frac{r}{1+r}$ measures the marginal propensity to consume and is known as the *perpetual income factor*. Note that, although consumer behavior is affected by uncertainty, only the *expected values* of y_t enter the definition of permanent income, while the variability of y_t does not affect consumption. We will return to this point.

2.3 Permanent Income and Shocks

We begin our investigation by starting with the case of quadratic utility when $\delta = r = 0$ and with an interval up to T. As we know, the Euler equation reduces to

$$E_t\left(c_{t+s}\right) = c_t, \tag{2.26}$$

or, equivalently,

$$c_{t+s} = c_t + \varepsilon_{t+s} \tag{2.27}$$

that is, the permanent income model implies that consumption follows a *random walk* process. The question is, what determines the ε_{t+s} shock in Eq. 2.27? Let us consider this aspect in detail.

When $\delta = r = 0$, the budget constraint in Eq. 2.21 reduces to

$$c_t = \frac{1}{T}\left[a_t + \sum_{s=0}^{T-t} \frac{E_t\left(y_{t+s}\right)}{(1+r)^s} \right], \tag{2.28}$$

since, at time t, Eq. 2.26 holds. Now, by the same reasoning, in $t+1$, the level of consumption c_{t+1} must be equal to $1/\left(T-1\right)$ of the total expected wealth over the remaining time horizon:

$$c_{t+1} = \frac{1}{T-1}\left[a_{t+1} + \sum_{s=1}^{T-t} \frac{E_{t+1}\left(y_{t+s}\right)}{(1+r)^s} \right]. \tag{2.29}$$

By substituting the budget constraint $a_{t+1} = a_t + y_t - c_t$ we can rewrite the above as

$$c_{t+1} = \frac{1}{T-1}\left[(a_t + y_t - c_t) + \sum_{s=1}^{T-t} \frac{E_{t+1}(y_{t+s})}{(1+r)^s}\right] \tag{2.30}$$

By adding and subtracting the value $\sum_{s=1}^{T-t} \frac{E_t(y_{t+s})}{(1+r)^s}$ from the latter equation, we derive the following:

$$c_{t+1} = \frac{1}{T-1}\left[(a_t + y_t - c_t) + \sum_{s=1}^{T-t} \frac{E_t(y_{t+s})}{(1+r)^s} + \left(\sum_{s=1}^{T-t} \frac{E_{t+1}(y_{t+s})}{(1+r)^s} - \sum_{s=1}^{T-t} \frac{E_t(y_{t+s})}{(1+r)^s}\right)\right]. \tag{2.31}$$

However, from Eq. 2.28 it follows that

$$c_{t+1} = \frac{1}{T-1}\left[c_t(T-1) + \left(\sum_{s=1}^{T-t} \frac{E_{t+1}(y_{t+s})}{(1+r)^s} - \sum_{s=1}^{T-t} \frac{E_t(y_{t+s})}{(1+r)^s}\right)\right]. \tag{2.32}$$

From this, with some algebra, we get

$$\Delta c_t = c_{t+1} - c_t = \frac{1}{T-1}\left(\sum_{s=1}^{T-t} \frac{E_{t+1}(y_{t+s})}{(1+r)^s} - \sum_{s=1}^{T-t} \frac{E_t(y_{t+s})}{(1+r)^s}\right). \tag{2.33}$$

The latter relationship states that the change in consumption Δc_t is equal to the change in estimated future income divided by the remaining time horizon. If, for example, the income an individual expects in $t+2$ is reduced, he or she will find it optimal to cope by compressing consumption and accumulating additional resources (by increasing his or her savings).

2.4 Uncertainty and Credit Rationing

In the previous chapter, we saw how the *liquidity constraint* affects the consumption dynamic over two periods. We can now generalize this result by considering the case of uncertain income and credit rationing, in a multi-period framework. It is worth noting that uncertain income *always* implies a liquidity constraint because any rational individual avoids of taking on debt in an amount greater than the minimum expected (future) income, even when capital markets are perfect, so as to mitigate the risk of default (Hayashi 1997).

We focus on the case where an individual lives for T periods and experiences uncertainty about his or her future income. This problem is examined in Zeldes (1989). We can write this as follows:

$$\begin{cases} V_t(a_t) = \max\limits_{c_t} E_t\left[\sum_{s=0}^{T-t} \frac{u(c_{t+s})}{(1+\delta)^s}\right] \\ \text{s.t. } a_{t+s+1} = (1+r)(a_{t+s} + y_{t+s} - c_{t+s}) \\ a_{t+s+1} \geqslant 0, \end{cases} \tag{2.34}$$

where the inequality $a_{t+s+1} \geqslant 0$ identifies the liquidity constraint on net wealth. Obviously, the transversality condition requires that $a_{T+1} = 0$. In the presence of a liquidity constraint, the HJB equation in $s = 0$ becomes

$$V_t(a_t) = \max_{c_t}\left\{u(c_t) + \frac{1}{1+\delta} E_t\left[V_{t+1}(a_{t+1})\right] + \mu(a_{t+1})\right\}, \tag{2.35}$$

where μ is the Lagrange multiplier associated with the constraint $a_{t+s+1} \geqslant 0$. The first-order condition with respect to c_t is

$$u'(c_t) = \frac{1+r}{1+\delta} E_t\left[V'_{t+1}(a_{t+1})\right] + \mu(1+r), \tag{2.36}$$

since, in $s = 0$, the budget constraint is $a_{t+1} = (1+r)(a_t + y_t - c_t)$. As explained above, to solve the HJB equation, we must differentiate Eq. 2.35 with respect to a_t

$$\begin{aligned} V'_t(a_t) = u'(c_t)\frac{\partial c_t}{\partial a_t} + \frac{1+r}{1+\delta} E_t\left[V'_{t+1}(a_{t+1})\right] - \\ - \frac{1+r}{1+\delta} E_t\left[V'_{t+1}(a_{t+1})\right]\frac{\partial c_t}{\partial a_t} + \mu(1+r)\left[1 - \frac{\partial c_t}{\partial a_t}\right] \end{aligned} . \tag{2.37}$$

That is,

$$\begin{aligned} V'_t(a_t) = \left[u'(c_t) - \frac{1+r}{1+\delta} E_t\left(V'_{t+1}(a_{t+1})\right) - \mu(1+r)\right]\frac{\partial c_t}{\partial a_t} + \\ + \frac{1+r}{1+\delta} E_t\left[V'_{t+1}(a_{t+1})\right] + \mu(1+r) \end{aligned} . \tag{2.38}$$

However, from the first-order condition, it follows that

$$u'(c_t) - \frac{1+r}{1+\delta} E_t\left[V'_{t+1}(a_{t+1})\right] - \mu(1+r) = 0. \tag{2.39}$$

Then, the envelope condition with respect to a_t implies

$$V'_t(a_t) = \frac{1+r}{1+\delta} E_t\left[V'_{t+1}(a_{t+1})\right] + \mu(1+r). \tag{2.40}$$

The latter equation solves Eq. 2.36. In fact, by substituting Eq. 2.40 into the Euler equation we obtain

$$u'(c_t) = V_t'(a_t). \tag{2.41}$$

That is, along the optimal trajectory, the marginal utility of consumption coincides with the marginal utility of wealth. This optimality relation must also be realized in $t+1$:

$$u'(c_{t+1}) = V_t'(a_{t+1}). \tag{2.42}$$

From this expression, by substituting in Eq. 2.36 we get

$$u'(c_t) = \frac{1+r}{1+\delta} E_t u'(c_{t+1}) + \mu(1+r) \geqslant \frac{1+r}{1+\delta} E_t u'(c_{t+1}), \tag{2.43}$$

which is the intertemporal optimality condition between two successive periods. It is valid as a strict equality only if $\mu = 0$. Note, if the constraint does not bind (*slack*), the Euler equation coincides with that in the unconstrained case. Again, the implications are the same as those of the two-period model with credit rationing: consumption grows over time because liquidity constraints have the effect of postponing individual consumption over time as a function of savings in previous periods. Finally, if $\mu > 0$, the marginal utility is no longer a martingale, even if $r = \delta$, since the liquidity constraint prevents continuous adjustment of consumption patterns.

2.5 Precautionary Saving

The *certainty equivalence model* in Sect. 2.2.3 allowed us to focus on some aspects of the consumer's intertemporal choice under uncertainty. In that framework, saving is just the difference between current income and its expected value. This removes the possibility of *extra* saving accompanying future income uncertainty.

Precautionary saving provides an explanation for the saving decisions documented in the empirical literature but not yet captured in the consumption models considered thus far: individuals with uncertain future income generally tend to save more in the present (Browning and Lusardi, 1996).

The idea that uncertainty has specific effects on saving can be traced back to the pioneering article by Leland (1968), who shows that with future uncertain income, there is an increase in saving if the third-order derivative of the utility function is positive. Leland referred to this *extra saving* as the precautionary demand for savings. A few years later, Sandmo (1970) and Rothschild and Stiglitz (1971) extended this original contribution to the case of random interest rates.

In the next section, we focus on this issue by considering the relationship between intertemporal consumption, risk aversion, and precautionary saving. To facilitate the analysis, we follow the original problem of Leland (1968) regarding saving decisions with uncertain income.

2.5.1 Risk Aversion and Precautionary Saving

It is useful to start from the definition of risk aversion. Traditionally, the degree of risk aversion is related to the *concavity* of the utility function measured by its first- and second-order derivatives. In a dynamic scenario, however, as we know, the optimal consumption policy is driven by the Euler equation. Therefore, to understand precautionary saving, we must focus on the first- and second-order derivatives of this condition. Precisely, we must consider the second- and third-order derivative of the utility function because, as we will explain, what now matters is the *convexity* of the Euler equation.

We must proceed carefully and introduce the concept of risk aversion starting from the static case. The following extension to the intertemporal context and the concept of precautionary savings will be more easily understood as a result.

Risk Aversion

We focus on the relationship between *risk aversion* and the utility function. It is well known that if an individual is risk-averse, then his or her utility function is *concave* (Hirshleifer and Riley 1992). By contrast, the utility function is *convex* for a risk-lover agent and *linear* for a risk-neutral one.

We explain the point by first assuming that the probability distribution p of a random income has only two pay-offs, measured in consumption goods, c_1 and c_2, with probabilities p_1 and $(1-p_1)$, respectively. The expected value is $E(c) = p_1c_1 + (1-p_1)c_2$. Formally, the utility representation for the risk-averse individual is

$$u\left[E(c)\right] > E\left[u(c)\right]. \tag{2.44}$$

That is, for a risk-averse agent, the utility of the expected value $E(c)$ is greater than the expected utility $E\left[u(c)\right]$ of consumption. Expanding Eq. 2.44, we get the expression

$$u\left[p_1c_1 + (1-p_1)c_2\right] > p_1u(c_1) + (1-p_1)u(c_2), \tag{2.45}$$

which is equivalent to the definition of *concavity* for the function $u(c)$, where $u'(c) > 0$ and $u''(c) < 0$.

The main implication of the above is that the first- and second-order derivatives of the utility function characterize the degree of risk aversion. Furthermore, by Eq. 2.45, a risk-averse individual will never accept a fair game when the *participation price* is equal to the *expected value* of the game itself since

$$E[u(c)] - u[E(c)] < 0; \tag{2.46}$$

in other words, a risk-averse individual, who takes part in a fair game, will always suffer a utility loss.

The graphical counterpart of this condition is depicted in Fig. 2.1. Every point D on the line $\overline{AB}$ can be seen as a linear combination of the two extreme points A and B weighted by p and $(1-p)$. Indeed, using the probabilities for c, we get

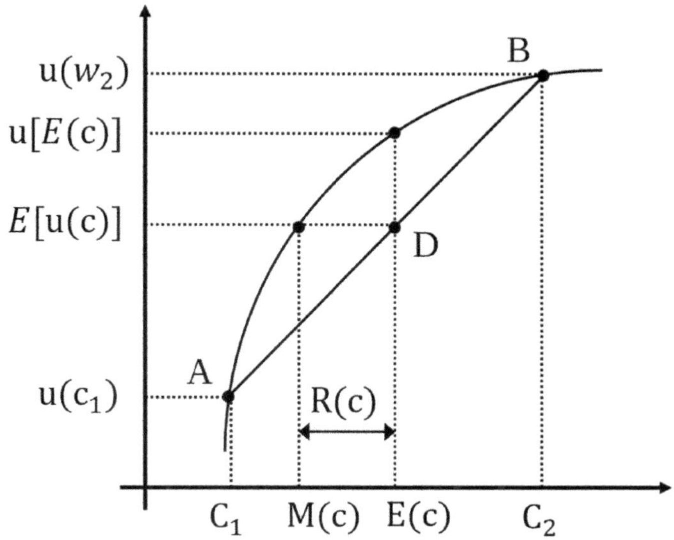

Fig. 2.1 Utility function and risk aversion

$$\begin{aligned} & p_1 \left[c_1, u(c_1)\right] + (1 - p_1) \left[c_2, u(c_2)\right] \\ & = (p_1 c_1 + (1 - p_1) c_2, p_1 u(c_1) + (1 - p_1) u(c_2)) \\ & = (E(c), E[u(c)]), \end{aligned} \tag{2.47}$$

which are the coordinates of the point D.

Note that the higher the curvature of $u(c)$, the greater the individual's risk aversion. However, since expected utility functions are not uniquely defined (are defined only up to linear transformations), rather than just the second-order derivative of $u(c)$, we need a constant measure with respect to these transformations. One such measure is the Arrow-Pratt measure of absolute risk aversion (λ), defined as $\lambda = -\frac{u''(c_t)}{u'(c_t)}$. An alternative index that does not depend on the unit of measure of consumption, income, or wealth is the coefficient of relative risk aversion, defined as $\rho = -\frac{u''(c_t)c_t}{u'(c_t)}$.

Finally, note that the concept of risk aversion is strictly related to the one of *risk premium*: the certain sum for which a risk-adverse agent is indifferent between either bearing the risk or receiving with certainty the expected consumption $E(c)$. It is defined as

$$u[E(c) - R(c)] = E[u(c)], \tag{2.48}$$

where $R(c)$ is the risk premium (see Fig. 2.1). Let's define $M(c)$ as the certainty equivalent. Then, the difference $R(c) = E(c) - M(c)$ provides the measure of the *risk premium* in terms of consumption goods, that is the (certain) sum an individual is willing to pay to be indifferent between bearing the "risk" or receiving the certain value $M(c)$. We can use $R(c)$ to compare individuals' risk aversion. More specifically, given the distribution p and the pay-offs, the greater the concavity of the utility function, the greater the risk premium an agent is willing to pay to avoid the risk. Therefore, risk-averse individuals will try to hedge against future risks by insuring themselves in the current period.

As we will show in the next section, precautionary saving is a form of self-insurance closely related to the concept of risk aversion and risk premium. However, in the intertemporal context, it is the *convexity* of the Euler equation that matters and not the *concavity* of the utility function.

Precautionary Saving

We begin by considering a two-period model where the second-period income is *uncertain*. We assume that the income in period $t+1$ can take one of two values: $y_{t+1,a}$ (in the best state) and $y_{t+1,b}$, (in the worst state), where $y_{t+1,a} > y_{t+1,b}$ with probabilities p and $(1-p)$, respectively. Finally, we assume that $y_t = E_t(y_{t+1})$, so that under certainty equivalence, consumers have no incentive to save for "hard times".

The consumer's problem can be stated as follows

$$\begin{cases} \max \quad u(c_t) + E_t\left[u(c_{t+1})\right] \\ \text{s.t.} \quad c_t + s_t = y_t \\ \quad c_{t+1} = \begin{cases} s_t + y_{t+1,a} & \text{with prob.} \quad p \\ s_t + y_{t+1,b} & \text{with prob.} \quad (1-p) \end{cases} \end{cases} \tag{2.49}$$

As usual, the first-order condition is the Euler equation

$$u'(c_t) = E_t\left[u'(c_{t+1})\right] \tag{2.50}$$

Substituting the budget constraint in Eq. 2.50 we get

$$u'(y_t - s_t) = E_t\left[u'(y_{t+1} + s_t)\right]. \tag{2.51}$$

Since by assumption $y_t = E_t(y_{t+1})$, Eq. 2.51 can be rewritten as

$$u'\left[E_t(y_{t+1}) - s_t\right] = E_t\left[u'(y_{t+1} + s_t)\right]. \tag{2.52}$$

As in the previous section the degree of risk aversion affects the savings decision. However, whereas in the static model we refer to the first- and second-order derivative of the utility function, now we focus on the sign of the derivatives of the Euler equation. Precisely, we refer to the second- and third-order derivatives of the utility function since the Euler equation already compares the first derivatives along the optimal consumption path.

Therefore, the value of saving s_t that satisfies Eq. 2.52 depends on the *convexity* of the *marginal utility* curve, that is on the sign of its third-order derivative $u'''(c)$.

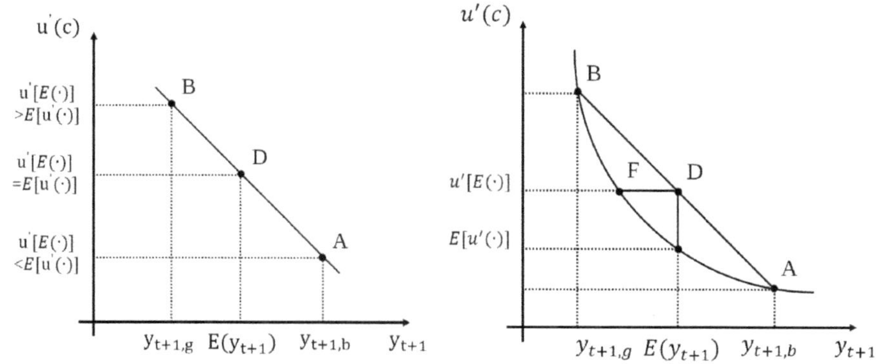

(a) $u(c_t) = ac_t - \frac{1}{2}bc_t^2$ (b) $u(c_t) = ln(c_t)$

Fig. 2.2 Linear and nonlinear marginal utility of consumption

Obviously, the extent of precautionary saving strictly depends on the form of the utility function and on its successive derivatives. We consider two examples (Fig. 2.2).

If the utility function has the quadratic form $u(c_t) = ac_t - \frac{1}{2}bc_t{}^2$, the consumer is risk averse. However, the marginal utility is linear (Fig. 2.2a) and the condition, Eq. 2.52, becomes

$$a - b\left[E_t(y_{t+1}) - s_t\right] = E_t\left[a - b(y_{t+1} + s_t)\right]. \tag{2.53}$$

Equation 2.53 is an equality between the expected value of marginal utility and the marginal utility of the expected value. Solving the expectation, we get

$$a - bE_t(y_{t+1}) - s_t = a - bE_t(y_{t+1}) + s_t, \tag{2.54}$$

and the optimal solution is

$$s_t{}^* = 0. \tag{2.55}$$

$$c_t{}^* = E_t\left(c_{t+1}{}^*\right). \tag{2.56}$$

The main consequence of this result is that, although consumers are risk-averse (i.e., $u'(c_t) > 0$ and $u''(c_t) < 0$), they do not react to risk. In other words, the consumer does not save in the first period, even if adverse shocks may decrease his or her future income. This happens because $u'''(c_t) = 0$. In this scenario, the intertemporal consumption plan depends only on the expected value of income and *not* on its variability. In sum, if marginal utility is linear, there will be no precautionary saving.

What does it happen if $u'''(c_t) > 0$, for example, in the case of the utility function $u(c_t) = ln(c_t)$? We know that for the logarithmic function

$$u'\left[E_t(y_{t+1})\right] < E_t\left[u'(y_{t+1})\right]; \tag{2.57}$$

that is, the marginal utility of the expected value is lower than the expected value of marginal utility.

Now, to satisfy the Euler equation, (2.51), consumers must reduce their current consumption, increasing it in the second period, so as to decrease his or her expected marginal utility in the second period, and increasing it in the first. Importantly, the *convexity* $u'(\cdot) = 1/x$ of the function $u(\cdot) = ln(c_t)$ implies that $u'''(\cdot) > 0$ (Fig. 2.2b). This means that the consumer saves in the present time to stabilize income in the second period. Note that, this only happens if the *third derivative* of the utility function is positive. In sum, only if $u'''(\cdot) > 0$ there is an incentive to put aside precautionary (extra) savings for the future.

Notice that, similar to the static model, even in the intertemporal context the concept of precautionary saving is related to the one of risk premium. Indeed, now, the segment FD in Fig. 2.2b measures the *extra saving* a consumer is willing to set aside to be indifferent between bearing the risk or benefit from the highest *marginal* utility associated to the corresponding certainty equivalent.

How is this incentive strong? One way to measure *absolute prudence* is the ratio between the third- and second-order derivatives of the utility function, $P(c_t) = -\frac{u'''}{u''}$ (Kimball 1990). Not surprisingly, absolute prudence is zero for the quadratic utility function and positive when the marginal utility is convex.

2.5.2 Euler Equation and Precautionary Saving

An explicit solution for consumption growth can be derived by considering a second-order approximation to the optimal consumption rule with constant interest rates. We rewrite the Euler equation by assuming a nonstochastic interest rate

$$u'(c_t) = \frac{1+r}{1+\delta} E_t\left[u'(c_{t+1})\right] \tag{2.58}$$

Following Blanchard and Mankiw (1988), Eq. 2.58 can be approximated by a second-order Taylor series expansion of $u'(c_{t+1})$ around c_t

$$u'(c_t) \cong \frac{1+r}{1+\delta} E_t\left[u'(c_t) + u''(c_t)(c_{t+1} - c_t) + \frac{1}{2}u'''(c_t)(c_{t+1} - c_t)^2\right] \tag{2.59}$$

Dividing Eq. 2.59 by c_t^2 we obtain

$$\begin{aligned}&\frac{u'(c_t)}{{c_t}^2} + \frac{u''(c_t)}{c_t} E_t\left(\frac{c_{t+1} - c_t}{c_t}\right)\\&+ \frac{1}{2}u'''(c_t) E_t\left(\frac{c_{t+1} - c_t}{c_t}\right)^2 \cong \frac{1+r}{1+\delta}\frac{u'(c_t)}{{c_t}^2}.\end{aligned} \tag{2.60}$$

Solving for the expected rate of growth of consumption, we obtain

$$E_t\left(\frac{c_{t+1}-c_t}{c_t}\right) \cong -\frac{u'(c_t)}{u''(c_t)\,c_t}\left[\frac{r-\delta}{1+r}\right] - \frac{u'''(c_t)\,c_t}{u''(c_t)}\frac{1}{2}E_t\left(\frac{c_{t+1}-c_t}{c_t}\right)^2, \tag{2.61}$$

or

$$E_t\left(\frac{c_{t+1}-c_t}{c_t}\right) \cong \text{EIS}\left[\frac{r-\delta}{1+r}\right] - AP(c_t)\frac{1}{2}E_t\left(\frac{c_{t+1}-c_t}{c_t}\right)^2, \tag{2.62}$$

where $\text{EIS} = -\frac{u'(c_t)}{u''(c_t)c_t}$ is the intertemporal elasticity of substitution, $AP(c_t) = -\frac{u'''(c_t)c_t}{u''(c_t)}$ is the coefficient of relative prudence of Kimball and $E_t\left(\frac{c_{t+1}-c_t}{c_t}\right)^2$ is a measure of the expected variability of consumption growth.

Essentially, Eq. 2.62 states that the variability of future consumption reduces the current rate of growth of consumption. Finally, the sensitivity of consumption growth to uncertainty depends on the coefficient of prudence. The empirical literature has estimated similar equations with income and/or consumption as the dependent variable, while trying to overcome the difficulties involved in obtaining a good measure of risk (Browning and Lusardi 1996; Lusardi 1998; Lugilde et al. 2019).

Example 2.1 We consider a numerical example to illustrate the difference in savings behavior in two cases: *certainty equivalence* and *precautionary saving*.

Suppose that the first-period income is $y_t = 10$ and the distribution of the second-period income y_{t+1} is as follows:

$$c_{t+1} = \begin{cases} y_{t+1,a} = 12 & \text{with prob. } 0.8 \\ y_{t+1,b} = 2 & \text{with prob. } 0.2 \end{cases} \tag{2.63}$$

It follows that $y_t = E_t(y_{t+1}) = 10$. In the case of complete markets, consumption in the second period is stabilized so that $c_t{}^* = c_{t+1,a}{}^* = c_{t+1,b}{}^* = 10$.

We now investigate the effect of *uncertainty* on saving. Assume that either (i) the utility function is *quadratic*, or (ii) the utility is *logarithmic*.

As shown above, with a quadratic utility function, marginal utility is $u'(c_t) = a - bc_t$. Therefore, $s_t{}^* = 0$ and $c_t{}^* = E_t(c_{t+1}{}^*) = 10$. Compared to the case of complete markets, under certainty equivalence, consumption is smoothed ex ante. Suppose now that the utility function is logarithmic, with marginal utility $u'(c_t) = \frac{1}{c_t}$. The first-order condition for an optimum is Eq. 2.51, that, given the assumptions on income distribution in Eq. 2.63, is defined as follows

$$\frac{1}{10-s_t} = \frac{0.8}{12+s_t} + \frac{0.2}{2+s_t} \tag{2.64}$$

Table 2.1 The effect of income uncertainty on consumption

Case	C_t	S_t	C_{t+1a}	C_{t+1b}
Complete markets	10	0	10	10
Quadratic function	10	0	12	2
Logarithmic function	8.54	1.46	13.46	15.46

We can easily verify that if $s_t = 0$, the first-order condition is violated. In fact,

$$u'(E_t y_{t+1}) = 0.1 < E_t u'(y_{t+1}) = 0.17. \tag{2.65}$$

To satisfy the condition, the marginal utility of consumption in the first period must be increased and the expectation for the second period must be reduced to close the gap between $u'\left[E_t\left(y_{t+1}\right)\right]$ and $E_t\left[u'(y_{t+1})\right]$. This is done by decreasing first-period consumption to 8.536 and increasing saving to $s_t{}^* = 1.464$. Table 2.1 summarizes the data.

2.6 Uncertain Returns: The Case of a Risky Asset

Let's now assume that the asset returns are uncertain. Suppose that there is a single risky asset a_t, which provides the uncertain return $(1 + z_t)$. In this case, the risk depends on the return $(1 + z_t)$ being known at time $t + 1$, after the consumption decision c_t has been made. For simplicity, we assume that returns follow a normal distribution and that labor income y_t is zero.

The problem is

$$\begin{cases} \max\limits_{c_t} E_t\left[\sum_{s=0}^{T-t} \frac{u(c_{t+s})}{(1+\delta)^s}\right] \\ \text{s.t. } a_{t+1} = (1 + z_t)(a_t - c_t) \\ E_t \prod_{s=0}^{T-t}(1 + z_{t+s})\, a_{T+1} = 0. \end{cases} \tag{2.66}$$

Recall that a_t is the state variable, while saving $s_t = (a_t - c_t)$ is the control variable. Finally, $E_t \prod_{s=0}^{T-t}(1 + z_{t+s})\, a_{T+1} = 0$ is the transversality condition.

2.6.1 Euler Equation

The HJB equation corresponding to Eq. 2.66 is

$$V_t(a_t) = \max_{c_t}\left\{u(c_t) + \frac{1}{1+\delta} E_t\left[V_{t+1}(a_{t+1})\right]\right\}; \tag{2.67}$$

that is,

$$V_t(a_t) = \max_{s_t} \left\{ u(c_t) + \frac{1}{1+\delta} E_t\left[V_{t+1}((1+z_t)s_t)\right] \right\}. \tag{2.68}$$

Since the budget constraint can be rewritten as $a_{t+1} = (1+z_t)s_t$, by differentiating with respect to s_t we obtain the first-order condition:

$$u'(c_t) = \frac{1}{1+\delta} E_t\left[V'_{t+1}((1+z_t)s_t)\right](1+z_t). \tag{2.69}$$

From the envelope condition, it follows that

$$V'_t(a_t) = u'(c_t). \tag{2.70}$$

Then, moving forward one period and substituting in Eq. 2.69, we obtain the corresponding *Euler equation*:

$$u'(c_t) = \frac{1+z_t}{1+\delta} E_t\left[u'(c_{t+1})\right]. \tag{2.71}$$

Let's assume that the utility function is $u(c_t) = \ln c_t$, so that the consumption function takes the following form:

$$c_t = \varphi a_t, \tag{2.72}$$

where φ is a constant to be determined. In other words, consumption depends only on individual wealth a_t, in a fixed proportion and equal to φ.

Equation 2.71 above, then becomes

$$\frac{1}{c_t} = \frac{1+z_t}{1+\delta} E_t\left[\frac{1}{c_{t+1}}\right]. \tag{2.73}$$

However, since $c_{t+1} = \varphi(1+z_t)(a_t - c_t)$, we can write the above as

$$\frac{1}{\varphi a_t} = \frac{1+z_t}{1+\delta} E_t\left[\frac{1}{\varphi(a_t - \varphi a_t)(1+z_t)}\right]. \tag{2.74}$$

Therefore, $\varphi = \left(1 - \frac{1}{1+\delta}\right)$. The consumption function then takes the form

$$c_t = \left(1 - \frac{1}{1+\delta}\right) a_t \tag{2.75}$$

2.7 Uncertain Returns with n-Risky Assets

We generalize the result above by assuming that there are n-risky securities and $1 + z_{it}$ denotes the return on asset i at time t. We assume that the different asset returns have zero correlation and are normally distributed. We simplify our problem by assuming that the planning horizon is infinite. The maximum problem takes the form

$$\begin{cases} \max_{c_t} E_t\left[\sum_{s=0}^{\infty} \frac{u(c_{t+s})}{(1+\delta)^s}\right] & \\ c_t + \sum_{i=1}^{n} s_{it} = a_t & \text{for } t \geq 0 \\ \text{s.t. } a_{t+1} = \sum_{i=1}^{n} s_{it}\left(1 + z_{it}\right) & \text{for } t \geq 0\,, \\ \lim_{t\to\infty} E_t\left(1 + z_t\right) a_t = 0 & \end{cases} \tag{2.76}$$

where a_0 is given and s_{it} is savings.

The corresponding HJB equation is

$$V_t\left(a_t\right) = \max_{s_{it}} \left\{u\left(c_t\right) + \frac{1}{1+\delta} E_t\left[V_{t+1}\left(a_{t+1}\right)\right]\right\}; \tag{2.77}$$

that is, substituting for c_t and a_{t+1},

$$V_t\left(a_t\right) = \max_{s_{it}} \left\{u\left(a_t - \sum_{i=1}^{n} s_{it}\right) + \frac{1}{1+\delta} E_t\left[V_{t+1}\left(\sum_{i=1}^{n} s_{it}\left(1 + z_{it}\right)\right)\right]\right\}. \tag{2.78}$$

From the latter equation, we obtain the first-order condition with respect to s_{it}:

$$u'\left(a_t - \sum_{i=1}^{n} s_{it}\right) = \frac{1 + z_{it}}{1+\delta} E_t V'_{t+1}\left(\sum_{i=1}^{n} s_{it}\left(1 + z_{it}\right)\right). \tag{2.79}$$

The envelope condition with respect to a_t is then

$$V'_t\left(a_t\right) = u'\left(a_t - \sum_{i=1}^{n} s_{it}\right). \tag{2.80}$$

Then, following the usual procedure, we derive the Euler equation from Eq. 2.79:

$$u'\left(a_t - \sum_{i=1}^{n} s_{it}\right) = \frac{1 + z_{it}}{1+\delta} E_t u'\left(a_{t+1} - \sum_{i=1}^{n} s_{it+1}\right). \tag{2.81}$$

To solve the Euler equation (2.81) we assume that the utility function is a CRRA, that is, $u\left(c_t\right) = \frac{c_t^{1-\gamma}}{1-\gamma}$ whose optimal saving function takes the form

$$s_{it} = w a_t \qquad \text{for each } i = 1 \ldots n, \tag{2.82}$$

with w being a constant to be determined. Substituting the first derivative of c_t with respect to the utility function, $u^{'}(c_t) = c_t^{-\gamma}$, into the Euler equation we have

$$c_t^{-\gamma} = \frac{1 + z_{it}}{1 + \delta} E_t \left(c_{t+1}^{-\gamma} \right). \tag{2.83}$$

From the initial conditions, we know that, for the nth title, the relation $c_t = a_t - s_{it}$ holds. Using Eq. 2.82 this can be rewritten as

$$c_t = (1 - w)\, a_t. \tag{2.84}$$

Substituting Eq. 2.84 into Eq. 2.83, we get

$$[(1 - w)\, a_t]^{-\gamma} = \frac{1 + z_{it}}{1 + \delta} E_t \left[(1 - w)\, a_{t+1} \right]^{-\gamma} \tag{2.85}$$

$$[(1 - w)\, a_t]^{-\gamma} = \frac{1 + z_{it}}{1 + \delta} E_t \left[(1 - w) \sum_{i=1}^{n} w a_{it}\, (1 + z_{it}) \right]^{-\gamma}. \tag{2.86}$$

Finally, recalling that $a_{t+1} = \sum_{i=1}^{n} s_{it}\, (1 + z_{it})$, dividing and simplifying, we get

$$w = \left(\frac{1 + z_{it}}{1 + \delta} \right)^{\frac{1}{\gamma}} E_t \left(\sum_{i=1}^{n} (1 + z_{it}) \right)^{-1}. \tag{2.87}$$

The economic implication of Eq. 2.87 is that with a CRRA function and a normal, independent distribution of returns, the consumer allocates a constant fraction of wealth to each asset. This fraction depends on the assumption of independence of the $1 + z_{it}$ returns over time and the normality of their distribution.

Appendix 1

Dynamic Programming Under Certainty

We conclude this chapter with two examples of dynamic programming *under certainty*. As will become clear, the procedure to solve the intertemporal consumption problem is similar to that applied *under uncertainty* (dealt with in the preceding sections), but with the simplification that in this case there are no expected values in the equations.

Example 2.2 (*The CRRA utility function*) Suppose that the consumer receives income in each period exclusively in the form of *returns* on wealth. Assume that labor income is zero, $y_t = 0$. Finally, the utility function is the CRRA $u(c_t) = \frac{c_t^{1-\gamma}}{1-\gamma}$. The HJB equation is

$$V_t(a_t) = \max_{c_t} \left\{ \frac{c_t^{1-\gamma}}{1-\gamma} + \frac{1}{1+\delta} \left[V_{t+1}\left((1+r)(a_t - c_t)\right) \right] \right\} \tag{A.1}$$

with the constraint

$$a_{t+1} = (1+r)(a_t - c_t). \tag{A.2}$$

The first-order condition is

$$c_t^{-\gamma} = \frac{1+r}{1+\delta} c_{t+1}^{-\gamma}. \tag{A.3}$$

Once again, we solve the first-order-difference equation by assuming that the *consumption function* takes the form:

$$c_t = \varphi a_t, \tag{A.4}$$

where φ is the constant to be determined. Note that in $t+1$ we have that $c_{t+1} = \varphi a_{t+1}$, or

$$c_{t+1} = \varphi (1+r)(a_t - c_t). \tag{A.5}$$

$$c_{t+1} = \varphi (1+r)(a_t - \varphi a_t). \tag{A.6}$$

Substituting for c_t and c_{t+1} in Eq. A.3 we get

$$(\varphi a_t)^{-\gamma} = \frac{1+r}{1+\delta} \left(\varphi (1+r)(a_t - \varphi a_t)\right)^{-\gamma}. \tag{A.7}$$

Simplifying this yields

$$\varphi = 1 - \frac{1}{\left(\frac{1}{1+\delta}\right)^{\frac{1}{\gamma}} (1+r)^{\frac{1-\gamma}{\gamma}}}. \tag{A.8}$$

Finally, by substituting φ into Eq. A.4 we obtain the consumption function,

$$c_t^*(a_t) = \left(1 - \frac{1}{\left(\frac{1}{1+\delta}\right)^{\frac{1}{\gamma}} (1+r)^{\frac{1-\gamma}{\gamma}}}\right) a_t, \tag{A.9}$$

where $c_t^*(a_t)$ denotes the optimal solution of the HJB equation.

From the optimal value of consumption, we can derive the form of the value function $V_t(a_t)$. We suppose it has the same form as the CRRA utility function and can be written as

$$V_t(a_t) = \phi \frac{a_t^{1-\gamma}}{1-\gamma}, \tag{A.10}$$

where ϕ is a constant to be determined. The corresponding HJB equation calculated in its optimal value $c_t^*(a_t)$ is

$$\phi \frac{a_t^{1-\gamma}}{1-\gamma} = \left\{ \frac{\left(\left(1-\frac{1}{A}\right)a_t\right)^{1-\gamma}}{1-\gamma} + \frac{1}{1+\delta}\phi \frac{a_{t+1}^{1-\gamma}}{1-\gamma} \right\}, \tag{A.11}$$

with $A = \left(\frac{1}{1+\delta}\right)^{\frac{1}{\gamma}} (1+r)^{\frac{1-\gamma}{\gamma}}$. From this expression, we can derive the value of the constant ϕ. Recalling the constraint expression and the solution, Eq. (A.9) for c_t, the preceding equation becomes

$$\phi \frac{a_t^{1-\gamma}}{1-\gamma} = \left\{ \frac{\left(\left(1-\frac{1}{A}\right)a_t\right)^{1-\gamma}}{1-\gamma} + \frac{1}{1+\delta}\phi \frac{\left[(1+r)\left(a_t - \left(1-\frac{1}{A}\right)a_t\right)\right]^{1-\gamma}}{1-\gamma} \right\}, \tag{A.12}$$

from which we obtain

$$\phi = \left[\frac{1-\frac{1}{A}}{1-\left(\frac{1+r}{1+\delta}\right)\left(\frac{\frac{1}{A}-\gamma}{1-\gamma}\right)} \right]. \tag{A.13}$$

Thus, the value function has the form

$$V_t(a_t) = \left[\frac{1-\frac{1}{A}}{1-\left(\frac{1+r}{1+\delta}\right)\left(\frac{\frac{1}{A}-\gamma}{1-\gamma}\right)} \right] \frac{a_t^{1-\gamma}}{1-\gamma}. \tag{A.14}$$

Appendix 2

Durable Goods

So far, we have assumed that utility depends only on consumption at time t. However, there are cases where it is not possible to verify the *intertemporal separability* of consumption. This is, for example, the case of *durables* or, more generally, goods that provide a flow of services over time (such as cars, computers or furniture).

The optimization problem is complicated by the presence of durable goods in the utility function because they introduce a time dependence between consumption decisions in different periods. Indeed, the utility of durables relates to the provision of services with effects on the utility of subsequent periods. There are thus two effects of consumption in each period: a *direct* effect on utility caused by the current consumption of a portion of durables and an *indirect* effect caused by the flow of services generated, in each period, by the future consumption of durables.

Following Mankiw (1982), we assume that the utility function has the form $u(c_t, D_t)$, where D_t denotes the gross stock of durables at time t. The accumulation of these goods is given by the expression

$$D_{t+1} = d_t + (1 - \eta) D_t, \tag{A.1}$$

where η is the rate of depletion and d_t is the flow of resources employed to accumulate new durables. Note that, if $\eta = 1$, the good is not durable and the model reduces to the standard form.

With durables the budget constraint becomes

$$a_{t+1} = (1 + r)(a_t + y_t - c_t - d_t). \tag{A.2}$$

Substituting for d_t in Eq. A.2, we obtain

$$a_{t+1} = (1 + r)\left[a_t + y_t - c_t - (D_{t+1} - (1 - \eta) D_t)\right]. \tag{A.3}$$

Again following Mankiw (1982), we consider the case in which the utility function depends only on durables ($c_t = 0$). The consumer's problem is expressed as follows:

$$\max_{D_t} \left[\sum_{s=0}^{T} \frac{u(D_{t+s})}{(1 + \delta)^s}\right], \tag{A.4}$$

and the consumer chooses D_t, given the constraint in Eq. A.3 and the transversality condition $(1 + r)^{-T} a_{T+1} = 0$.

The corresponding HJB equation is

$$V_t(a_t) = \max_{D_t} \left\{u(D_t) + \frac{1}{1 + \delta}\left[V_{t+1}(a_{t+1})\right]\right\} \tag{A.5}$$

and substituting the constraint in Eq. A.3 we get

$$V_t(a_t) = \max_{D_t} \left\{u(D_t) + \frac{1}{1 + \delta} V_{t+1}(1 + r)\left[a_t + y_t - c_t - D_{t+1} + (1 - \eta) D_t\right]\right\}. \tag{A.6}$$

The first-order condition becomes

$$D_t : u'(D_t) + \frac{1 + r}{1 + \delta}(1 - \eta) V'_{t+1}(a_{t+1}) = 0, \tag{A.7}$$

which, as in the standard case, requires the application of the envelope theorem to be solved. Therefore, differentiating with respect to a_t we obtain

$$V_t'(a_t) = u'(D_t)\frac{\partial D_t}{\partial a_t} + \frac{1+r}{1+\delta}\left[V_{t+1}'(a_{t+1})\right] + \frac{1+r}{1+\delta}\left[V_{t+1}'(a_{t+1})\right]\frac{\partial D_t}{\partial a_t}(1-\eta). \tag{A.8}$$

that is,

$$V_t'(a_t) = \left[u'(D_t) + \frac{1+r}{1+\delta}\left(V_{t+1}'(a_{t+1})\right)(1-\eta)\right]\frac{\partial D_t}{\partial a_t} + \frac{1+r}{1+\delta}\left[V_{t+1}'(a_{t+1})\right]. \tag{A.9}$$

So, the first-order condition remains identical to that of the standard model:

$$V_t'(a_t) = \frac{1+r}{1+\delta}\left[V_{t+1}'(a_{t+1})\right]. \tag{A.10}$$

Equation A.10 can be rewritten as

$$u'(D_t) = \frac{r+\eta}{1+r}V_t'(a_t), \tag{A.11}$$

where $\frac{r+\eta}{1+r}$ is the discounted value of the user cost of durables. The user cost of durables is equal to the opportunity cost of the intertemporal value of consumption (r) plus the depreciation rate η of that good. From the previous two equations and the first-order condition, we can finally obtain the condition

$$u'(D_t) = \frac{1+r}{1+\delta}u'(D_{t+1}), \tag{A.12}$$

It is equal to the Euler equation for non-durable goods.

Appendix 3

HARA Functions

The utility functions used so far allow us to solve the consumer's problem analytically. These functions derive from a class of functions known as hyperbolic absolute risk aversion (HARA) functions.

A function is of HARA-type if its absolute risk aversion is defined as $\left(-\frac{u''(y_s)}{u'(y_s)} = \lambda\right)$, and its inverse $\left(\tau = \frac{1}{\lambda}\right)$, measuring the degree of risk tolerance, is linear in y_s. The HARA function takes the form

$$u(y_s) = a\left(\eta + \frac{y_s}{\gamma}\right)^{1-\gamma}, \tag{A.1}$$

where y_s is a variable (e.g., income), a, η, γ are parameters and $\left(\eta + \frac{y_s}{\gamma}\right) > 0$.

We compute the first- and second-order derivatives and the *absolute risk aversion coefficient* λ

$$u'(y_s) = \frac{a(1-\gamma)}{\gamma}\left(\eta + \frac{y_s}{\gamma}\right)^{-\gamma}, \tag{A.2}$$

$$u''(y_s) = -\frac{a(1-\gamma)}{\gamma}\left(\eta + \frac{y_s}{\gamma}\right)^{-\gamma-1}. \tag{A.3}$$

We assume that $-\frac{a(1-\gamma)}{\gamma} > 0$. Recalling that $\lambda = -\frac{u''(y_s)}{u'(y_s)}$, we get

$$\lambda = \left(\eta + \frac{y_s}{\gamma}\right)^{-1} \tag{A.4}$$

and

$$\rho = y_s\left(\eta + \frac{y_s}{\gamma}\right)^{-1}, \tag{A.5}$$

since $\rho = -\frac{u''(y_s)}{u'(y_s)} y_s$. The utility functions that have such a λ belong to the HARA class because they have a hyperbolic absolute risk aversion coefficient.

The *coefficient of risk tolerance* is $\tau = -\frac{u'(y_s)}{u''(y_s)}$. Since τ is the reciprocal of the hyperbolic function λ, for the HARA functions we have that

$$\tau = \eta + \frac{y_s}{\gamma}. \tag{A.6}$$

This coefficient is often written as

$$\tau = c + d y_s, \tag{A.7}$$

where $c = \eta$ and $d = \frac{1}{\gamma}$. Hence, HARA functions exhibit linear risk tolerance.

From the HARA function, we can derive the utility functions encountered in our analysis so far, namely (i) the CRRA function, (ii) the constant absolute risk aversion (CARA) function, and (iii) the quadratic function.

CRRA Function

If $\eta = 0$, then λ equals γ, which must be assumed to be non-negative. The case $\eta = 0$ corresponds to the situation where the relative risk aversion is independent of wealth: the relative risk premium that an individual is ready to pay to escape a multiplicative risk does not depend on wealth. Using the Eq. A.3, we can select a to normalize $u'(1) = 1$. This implies that this set of utility functions is defined as

$$u'(y_s) = y_s^{-\gamma}, \tag{A.8}$$

which means

$$u(y_s) = \begin{cases} \frac{{y_s}^{1-\gamma}}{1-\gamma} & \text{if} \quad \gamma \neq 1 \\ ln(y_s) & \text{if} \quad \gamma = 1 \end{cases} \tag{A.9}$$

It is worth noting that the asymptotic properties of these functions are quite different, depending upon whether relative risk aversion is larger or smaller than unity. Indeed, when γ is less than unity, utility goes from 0 to ∞ as wealth goes from zero to infinity. But when γ is greater than 1, the utility goes from $-\infty$ to zero.

CARA Functions

Let us recall the HARA function in Eq. A.4. We can observe that the absolute risk aversion λ is independent of y_s if $\gamma \to +\infty$. Indeed, in this case we get $\lambda = \frac{1}{\eta}$. It can be easily verified that the set of functions satisfying the differential equation $-u''(y_s) = \lambda u'(y_s)$ corresponds to

$$u(y_s) = -\frac{exp(-\lambda y_s)}{\lambda}, \tag{A.10}$$

which is a CARA function.

Quadratic Function

Let us start again with the function

$$u(y_s) = a\left(\eta + \frac{y_s}{\gamma}\right)^{1-\gamma}. \tag{A.11}$$

If $\gamma = -1$:

$$\begin{aligned} &\Rightarrow u(y_s) = a(\eta - y_s)^2 \\ &= a\left(\eta^2 + {y_s}^2 - 2\eta y_s\right) \\ &= \left(a\eta^2 + a{y_s}^2 - 2a\eta y_s\right). \end{aligned} \tag{A.12}$$

Further, suppose for simplicity that

$$\alpha = a\eta^2 \quad \text{and} \quad \beta = 2a\eta. \tag{A.13}$$

The function $u(y_s)$ can be rewritten as

$$u(y_s) = \alpha + a{y_s}^2 - \beta y_s. \tag{A.14}$$

If we set

$$\begin{aligned}\alpha &= 0\\ a &= -\frac{\epsilon}{2}\\ \beta &= 1,\end{aligned} \tag{A.15}$$

the function becomes

$$u\left(y_s\right) = y_s - \frac{\epsilon}{2}y_s^2 \tag{A.16}$$

We verify that this quadratic equation is a HARA function by computing the corresponding risk tolerance index. We obtain

$$\begin{aligned}u'\left(y_s\right) &= 1 - \epsilon y_s\\ u''\left(y_s\right) &= -\epsilon\\ \lambda &= -\frac{u''\left(y_s\right)}{u'\left(y_s\right)} = \frac{\epsilon}{1-\epsilon y_s};\end{aligned} \tag{A.17}$$

that is,

$$\tau = \frac{1}{\lambda} = \frac{\epsilon}{1-\epsilon y_s}. \tag{A.18}$$

We often encounter these functions in the consumption models of economics and finance presented in the book.

Further Reading.
Many textbooks address discrete-time optimization under uncertainty through the application of consumption theory. For an in-depth treatment of concepts from probability theory, stochastic difference equations, and stochastic dynamic programming in discrete time, we suggest Miao (2020) (Chaps. 2–8). Well-known expositions on the subject with applications to economics and finance are those by Malliaris and Brock (1977) (Chaps. 1–3), Dixit (1993) (Chaps. 1–4) and Ljungqvist and Sargent (2018) (Parts I and II). Other remarkable textbooks include *Models For Dynamic Macroeconomics* by Bagliano and Bertola (2004) (Chap. 1) and *The Economics of Consumption* by Jappelli and Pistaferri (2017) (Chaps. 4 and 11). Additional applications of stochastic dynamic programming to general economic problems can be found in Stokey and Lucas (1989) (Chap. 10). Concepts and solution methods for finite-horizon control problems under uncertainty in discrete time are also thoroughly discussed in an article by Runggaldier (1998) and the book by Chiang (1992).

References

Bagliano, F.-C., & Bertola, G. (2004). Models for dynamic macroeco- nomics. Oxford University Press.

Bellman, R. E. (1957). Economic growth. Princeton University Press.

Blanchard, O. J., & Mankiw, G. N. (1988). Consumption: Beyond cer- tainty equivalence. The American Economic Review, 78 (2), 173- 177.

Browning, M., & Lusardi, A. (1996). Household saving: Micro theories and micro facts. Journal of Economic literature, 34 (4), 1797- 1855.

Chiang, A. (1992). Elements of dynamic optimization. Waveland Press, Inc.

Dixit, A. K. (1993). The art of smooth pasting. first edition. Taylor & Francis.

Hall, R. E. (1978). Stochastic implications of the life cycle-permanent in- come hypothesis: Theory and evidence. Journal of political econ- omy, 86 (6), 971-987.

Hayashi, F. (1997). Understanding saving: Evidence from the united states and japan. MIT University Press.

Hirshleifer, J., & Riley, J. G. (1992). The analytics of uncertainty and information. Cambridge University Press.

Jappelli, T., & Pistaferri, L. (2017). The economics of consumption: Theory and evidence. Oxford University Press.

Kimball, M. S. (1990). Precautionary saving and the marginal propen- sity to consume.

Leland, H. E. (1968). Saving and uncertainty: The precautionary de- mand for saving. The Quarterly Journal of Economics, 82 (3), 465-473.

Ljungqvist, L., & Sargent, T. J. (2018). Recursive macroeconomic the- ory. MIT University Press.

Lugilde, A., Bande, R., & Riveiro, D. (2019). Precautionary saving: A review of the empirical literature. Journal of Economic Surveys, 33 (2), 481-515.

Lusardi, A. (1998). On the importance of the precautionary saving mo- tive. The American Economic Review, 88 (2), 449-453.

Malliaris, A. G., & Brock, W. A. (1977). Stochastic methods in eco- nomics and finance. North Holland.

Mankiw, G. N. (1982). Hall's consumption hypothesis and durable goods. Journal of Monetary Economics, 10 (3), 417-425.

Miao, J. (2020). Economic dynamics in discrete time. MIT University Press.

Rothschild, M., & Stiglitz, J. E. (1971). Increasing risk ii: Its economic consequences. Journal of Economic theory, 3 (1), 66-84.

Runggaldier, W. J. (1998). Concepts and methods for discrete and con- tinuous time control under uncertainty. Insurance: Mathematics and Economics, 22 (1), 25-39.

Sandmo, A. (1970). The effect of uncertainty on saving decisions. The review of economic studies, 37 (3), 353-360.

Stokey, N. L., & Lucas, R. E. (1989). Recursive methods in economic dynamics. Harvard University Press.

Zeldes, S. P. (1989). Consumption and liquidity constraints: An empir- ical investigation. Journal of political economy, 97 (2), 305-346. .

Consumption and Portfolio Rules 3

- In this chapter, we study the portfolio choices of *risk-averse* individuals.
- We assume that n-risky assets and one *risk-free* asset exist.
- From the Euler equation, we derive the equilibrium relation known as *consumption capital asset pricing model (CCAPM)*.
- This relation expresses the *risk premium* as a function of the negative correlation between consumption and expected returns.
- From the CCAPM model, we then derive the *security market line* (SML).
- The SML allows *unsystematic* or diversifiable risk to be distinguished from *systematic* risk as measured by the β index.
- The assumptions of *efficient markets* and *random walk* dynamics of stock prices are consistent with SML and the absence of arbitrage.
- Finally, we investigate anomalies, including the *equity premium* and *risk-free* rate puzzles.

3.1 Introduction

In our explorations thus far, we have assumed that uncertainty in consumption decisions is the result of either income uncertainty or variability in financial returns. We now consider the *general* case in which y_t and z_t fluctuate *randomly together*.

We assume that the individual is *uncertain* about the evolution of his or her income y_t and returns z_t on assets. Each individual must choose the optimal consumption and savings allocations *simultaneously* to maximize his or her expected utility. The outcome of this process depends on the factors influencing the intertemporal choice, including the individual's *degree of risk aversion* and the *covariance* between income and financial returns.

G. Travaglini et al., *Notes on Consumption Theory*, Classroom Companion: Economics,
https://doi.org/10.1007/978-3-031-54986-1_3

With this extension, the model has additional implications, specifically for the dynamic trends in consumption and rates of return. As we will see, the model is the basis for the *asset pricing* relationship known as the *consumption capital asset pricing model* (CCAPM). This type of problem was originally discussed by Samuelson (1969) and Merton (1969, 1971).

3.2 Consumption and Portfolio Rules

Consider the consumption problem over a finite horizon:

$$V_t(a_t) = \max_{c_t} E_t\left[\sum_{s=0}^{T-t} \frac{u(c_{t+s})}{(1+\delta)^s}\right], \tag{3.1}$$

where $u(c_t)$ is, as usual, the separable and additive utility function, and δ is the rate of time preference. Compared to the problems in Chap. 2, here, both income and returns are uncertain. Suppose that only two financial assets exist: a *risky* asset with an uncertain return $(1+z_t) = Z_t$, and a *riskless* asset, with a certain return $(1+r_t) = R_t$.

The intertemporal budget constraint can be written as

$$a_{t+s+1} = (a_{t+s} + y_{t+s} - c_{t+s})\left[R_{t+s}w + Z_{t+s}(1-w)\right], \tag{3.2}$$

where w is the share of the financial portfolio invested in the non-risky asset.

The corresponding HJB equation is

$$V_t(a_t) = \max_{c_t, w_t}\left\{u(c_t) + \frac{1}{1+\delta}E_t\left[V_{t+1}(a_{t+1})\right]\right\}. \tag{3.3}$$

Substituting the constraint in Eq. 3.2, it becomes

$$V_t(a_t) = \max_{c_t, w_t}\left\{u(c_t) + \frac{1}{1+\delta}E_t\left[V_{t+1}((a_t + y_t - c_t)[R_t w + Z_t(1-w)])\right]\right\}. \tag{3.4}$$

The solution of this problem requires two first-order conditions, one for c_t and the other for the share w:

$$c_t : u'(c_t) = \frac{1}{1+\delta}E_t\left[V'_{t+1}(a_{t+1})(R_t w + Z_t(1-w))\right], \tag{3.5}$$

and

$$w : 0 = \frac{1}{1+\delta}E_t\left[V'_{t+1}(a_{t+1})(R_t - Z_t)\right]. \tag{3.6}$$

We solve these two Euler equations by exploiting the envelope relationship between $V_t'(a_t)$ and $u'(c_t)$ along the optimal path:

$$V_t'(a_t) = \frac{1}{1+\delta} E_t \left[V_{t+1}'(a_{t+1}) (R_t w + Z_t (1-w)) \right]. \tag{3.7}$$

Then, by substituting Eq. 3.7 in Eq. 3.5, moving forward one period, and substituting in the two first-order conditions, we get

$$u'(c_t) = \frac{1}{1+\delta} E_t \left[(R_t w + Z_t (1-w)) \right] u'(c_{t+1}) \tag{3.8}$$

and

$$E_t \left[u'(c_{t+1}) R_t \right] = E_t \left[u'(c_{t+1}) Z_t \right] \tag{3.9}$$

Equivalently, we can substitute Eq. 3.5 directly into Eq. 3.6, deriving the following conditions

$$u'(c_t) = \frac{R_t}{1+\delta} E_t \left[u'(c_{t+1}) \right] \tag{3.10}$$

and

$$u'(c_t) = \frac{1}{1+\delta} E_t \left[Z_t u'(c_{t+1}) \right] \tag{3.11}$$

The economic meaning of Eqs. 3.10 and 3.11 is as follows: in presence of uncertainty about income and returns, the Euler equation must be satisfied for both riskless *and* risky assets.

We explain this point by focusing on Eq. 3.11. It claims that a decrease in consumption at time t equal to dc_t implies an increase in $u'(c_t)$. But, Eq. 3.11 also states that dc_t is now invested in risky assets with returns of Z_t and the marginal utility in $t+1$ is equal to $u'(c_{t+1})$. Obviously, the Euler equation holds only in expected value since it depends on $E_t \left[Z_t u'(c_{t+1}) \right]$.

Similarly, uncertainty affects the optimal condition in Eq. 3.10, whereby a change in consumption implies that the marginal utility in t must equal the discounted value of expected marginal utility in the next period.

In essence, Eqs. 3.10 and 3.11 define a set of *restrictions* on the dynamic process of consumption and portfolio rules between risky and non-risky assets along the optimal consumption path, respectively. They can be interpreted in one of two ways, as imposing restrictions on:

1. *consumption*, given the stochastic process that determines the asset returns, or
2. on *asset returns*, given the optimal consumption process.

In the first case, we study the dynamics of *consumption*, as reflected in the certainty equivalence model studied above. In the second case, we consider the dynamic relationship between *certain* and *uncertain* returns, given the utility function, level of risk aversion, and covariance between returns and consumption. In this latter case, we obtain the asset pricing rule known as CCAPM. We derive this relationship in the next section.

3.3 The Consumption Capital Asset Pricing Model

We derive the CCAPM by first noting that with n-risky assets, the optimal condition in Eq. 3.11 must hold for any asset. Hence, it can be written that for each risky asset $i = 1, 2...n$, it must be verified that

$$u'(c_t) = \frac{1}{1+\delta} E_t \left[Z_{it} u'(c_{t+1}) \right]. \tag{3.12}$$

Equalizing Eqs. 3.10 and 3.12 gives the expression:

$$E_t \left[u'(c_{t+1}) (z_{it} - r_t) \right] = 0 \tag{3.13}$$

This condition can be rewritten, keeping in mind that the expected value of the product of two (normally distributed) stochastic variables x and y is equal to $E(xy) = E(x)E(y) + Cov(x, y)$. By exploiting this rule, the optimal condition in Eq. 3.13 becomes

$$E_t u'(c_{t+1}) E_t (z_{it} - r_t) + Cov \left[u'(c_{t+1}), z_{it} \right] = 0, \tag{3.14}$$

where $Cov \left[u'(c_{t+1}), z_{it} \right]$ measures the covariance between the two variables. Therefore, the excess of the *expected* return, z_{it}, of the risky asset over the *riskless* rate r_t can be expressed as follows:

$$E_t (z_{it}) - r_t = - \frac{Cov \left[u'(c_{t+1}), z_{it} \right]}{E_t u'(c_{t+1})}. \tag{3.15}$$

This is the fundamental CCAPM equation. It states that a risk-averse individual requires the *risk premium*:

$$E_t (z_{it}) - r_t. \tag{3.16}$$

to hold a risky asset. This difference measures the extent of the risky return that must exceed the one of the riskless asset in order to motivate a risk-averse agent to hold a risky asset.[1] Notably, from Eq. 3.15, this value is proportional to the ratio between the covariance $Cov \left[u'(c_{t+1}), z_{it} \right]$ of the marginal utility of consumption with the uncertain return and the expected value of the marginal utility.

[1] Quoting Arrow (1996, p. 155), if the expected excess return is positive, it is a favorable gamble, and "a risk averter...always takes some part of a favorable gamble".

What does this mean? By assumption the denominator of the ratio is positive, but, in contrat, the numerator is negative. Indeed, the risky return always covaries *negatively* with the marginal utility (positively with consumption) so *destabilizing* consumption. Hence, since $Cov\left[u'(c_{t+1}) z_{it}\right] < 0$ the right-hand side of Eq. 3.15 is positive, given the minus sign multiplying the ratio.

As a further explanation of this condition, note that if an asset provides a low return when consumption is low (and the corresponding marginal utility is high) then the covariance between the return and the marginal utility of consumption is negative. In other words, the asset is risky precisely because it does not offer high returns when these would be most appreciated by the consumer. Therefore, with decreasing marginal utility, the risk-adverse individuals will be willing to hold risky assets *only if* they provide a positive risk premium in return.

3.3.1 The Security Market Line

It is clear from Eq. 3.15 that the risk premium depends on the utility function and the degree of risk aversion: the more convex the function $u(c_t)$, the higher the ratio $-\frac{Cov[u'(c_{t+1}), z_{it}]}{E_t u'(c_{t+1})}$. This property is actually unattractive to investors because it makes the risk premium depend directly on the shape of the utility function. Thus, it seems that individuals with different degrees of risk aversion can only transact for a risky asset fortuitously.

However, from the relationships considered above, it is possible to derive an equilibrium condition that represents the entire financial market, and that is independent of $u(c_t)$. This relationship is known as *security market line* (SML).

We obtain the SML by first assuming that m is a portfolio representative of the complete financial market and for which the risky return z_{mt} is negatively correlated with $u'(c_{t+1})$. We refer to m as the market portfolio. We can write

$$u'(c_{t+1}) = -\mu z_{mt}, \tag{3.17}$$

where μ measures the correlation between the rate of risky return, z_{mt}, and the marginal utility of consumption. Given the portfolio m, Eq. 3.15 can then be written as

$$E_t(z_{mt}) = r_t - \frac{Cov[-\mu z_{mt}, z_{mt}]}{E_t u'(c_{t+1})}. \tag{3.18}$$

Noting that since $Cov[z_{mt}, z_{mt}] = var(z_{mt})$, the previous relationship becomes

$$E_t(z_{mt}) = r_t - \frac{\mu var(z_{mt})}{E_t u'(c_{t+1})}, \tag{3.19}$$

where the term $Cov[z_{mt}, z_{mt}]$ denotes the variance of z_{mt}. From Eq. 3.18 we derive

$$\frac{1}{E_t u'(c_{t+1})} = \frac{E_t(z_{mt}) - r_t}{\mu var(z_{mt})}. \tag{3.20}$$

Finally, substituting Eqs. 3.17 and 3.20 into Eq. 3.15 we obtain

$$E_t(z_{it}) = r_t + [E_t(z_{mt}) - r_t]\beta_i, \tag{3.21}$$

with $\beta_i = Cov[z_{it}, z_{mt}]/var(z_{mt})$, and $E_t(z_{mt}) - r_t > 0$. Equation 3.21 is known as the SML. This linear relationship states that, given the market portfolio m, the risk premium β_i of any risky asset i is given by the ratio between its covariance with m and its variance

$$\beta_i = \frac{\sigma_{im}}{\sigma^2{}_m} \tag{3.22}$$

It implies that although a security i may be risky on its own, it may be *less* risky within a portfolio of risky assets if its expected return is negatively correlated with the expected return of m. Further, Eq. 3.21 has a special property: it has the advantage of making the asset pricing relationship *independent* of the utility function and degree of risk aversion.

Thus, for an asset i traded in the market, the appropriate measure of risk is not its variance, σ_i, but *only* that part of the systematic risk β_i related to the market portfolio, m. More precisely:

1. σ_i measures the *overall* risk of an asset;
2. $\beta_i = \sigma_{im}/\sigma^2{}_m$ represents the *residual* risk of an asset when compared with other assets in the portfolio; β_i is thus referred to as the *systematic component* of risk.
3. The difference between (i) and (ii),

$$[\sigma_i - \beta_i], \tag{3.23}$$

measures that part of the risk that cannot be eliminated through diversification. This is the *non-systematic* risk component.

Figure 3.1 is a graphical representation of the SML.

3.3.1.1 Beta and Systematic Risk

Equation 3.21 can be rewritten in an alternative form by substituting the definitions $Cov[z_{it}, z_{mt}] = \rho_{im}\sigma_i m$ and $var(z_{mt}) = \sigma_m^2$, where $-1 \leq \rho_{im} \leq 1$ is the index of correlation between the *risky* asset i and the *market* portfolio. The SML then becomes

$$E_t(z_{it}) = r_t + \lambda\hat{\beta}_i, \tag{3.24}$$

where $\hat{\beta}_i = \rho_{im}\sigma_i$, while $\lambda = \frac{[E_t(z_{mt}) - r_t]}{\sigma_m}$.

In this formulation, the risk premium is broken down into two components. The first $\frac{[E_t(z_{mt}) - r_t]}{\sigma_m}$, measures the risk premium per *unit* of risk. It states the extent to which the expected return on asset i should increase with each one-unit increase of that asset. Since this ratio is not affected by the convexity of the utility function, the

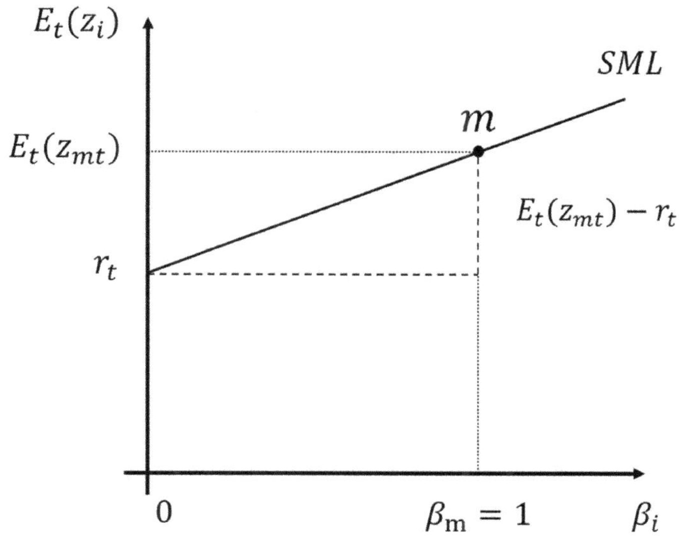

Fig. 3.1 The security market line (SML)

unit risk premium defines the unitary cost of risk that the market attaches to the risky asset. This index is known in the literature as the *Sharpe ratio* (Sharpe 1966).

The second component of the risk premium is $\hat{\beta}_i$, which accounts for the systematic unitary cost of risk attributed to asset i when compared with the risk of the market portfolio, m. For example, if $\rho_{im} = 0$, then $\hat{\beta}_i = 0$. Also, as before, this component does not depend on the characteristics of the utility function and is an "objective" measure of the systematic risk of the asset relative to the portfolio m.

3.4 SML and Efficient Markets

The relationship between the SML and the concept of *efficient financial markets* is an important one.

The *efficient market hypothesis* (EMH) is the core of modern financial market theory. According to the definition in Fama (1970), a financial market is efficient if the price of the assets fully reflects the information available in the market. The importance of this definition is indisputable: the EMH excludes the possibility of implementing investment strategies that, in a systematic manner, ensure returns in excess of those compatible with the SML. In other words, systematic *extra* returns are not compatible with the EMH, and *arbitrage* is the process by which agents adjust the value of returns to be consistent with the SML.

Hence, under the assumption of an efficient market an investor cannot obtain returns *systematically* higher than those consistent with the market equilibrium. In fact, any excess demand (positive or negative) is transmitted to the market by price changes, making any effort by investors to "beat" the market vain. In short, price changes are the only data that capture the information dispersed in the market and underlying portfolio decisions.

During the 1970s, quantitative investigations to test the EMH provided a number of robust empirical results, and there appeared to be consensus on the idea that financial markets were indeed efficient. Quoting Jensen (1978, p. 95): "There is no other proposition in economics which has more solid empirical evidence supporting it than the Efficient Market." This attitude implied that financial economics was moving toward the establishment of *asset pricing* models based on the concept of *efficiency* and the absence of opportunities for arbitrage. The CCAPM studied above and, more generally, the set of mean-variance models represent the most obvious extension of this view.

However, empirical research has shown that even findings around which there is consensus can contain controversial aspects that may reduce the EMH's ability to interpret stylized facts. In particular, arbitrage has proven to be less robust than claimed in previous analyses. For example, the stock market crash of 1987, the speculative bubble of 1995–2000, and the international financial crisis of 2008–2010 seem incompatible with the EMH. The question we need to ask, therefore, is whether financial markets are able to operate efficiently or, rather, whether "systematic and significant deviations from efficiency are expected to persist for long periods of time" (Shleifer 2000, p. 2).

Therefore, in this section, we introduce the essential elements that underlie the EMH according to the intertemporal consumption theory considered so far. We describe the theoretical models and empirical analyses on which the hypothesis is based. Then, we review the limitations of the models and new developments in financial market theory.

3.4.1 The Theoretical Foundations of the EMH

The EMH is based on *three* assumptions (in order of stringency):

1. In its narrowest formulation, the EMH assumes that investors are rational, follow optimizing behaviors and thus price the assets on the value of their fundamentals (expected profit, risk, rate of return, etc.);
2. In a less stringent formulation, the EMH assumes that a proportion of investors behave irrationally but that, on average, their inefficient strategies cancel out, not affecting the equilibrium price of assets;
3. Finally, in the most general form of the EMH, it is assumed that many investors are irrational. However, the presence of a rational few is sufficient for arbitrage strategies to prevail, such that returns and asset prices return to their equilibrium values.

We start with the first assumption. If all investors are rational, asset prices and returns are determined by the principle of profit and utility maximization already expressed by *Fisher's theorem*. In the present context, when investors receive *new* information about a particular asset, they adjust their portfolio composition, leading to a change in prices. Thus, the price of a security reflects the totality of information in the market at any given time, influencing portfolio choices (Samuelson 1969).

However, investors are not always rational, and their expectations are heterogeneous. In this second scenario, the EMH continues to emerge as a robust hypothesis. Indeed, in the presence of an abundance of investors, inefficient trades tend to cancel out, and the price of financial assets converges to the equilibrium price expressed by the *SML*. Clearly, this result is based on the assumption that individual strategies are uncorrelated and, therefore, there are no systematic distorting effects on the equilibrium value of expected returns.

This leads to the third form. When the correlation between individual decisions *affects prices*, it creates opportunities for arbitrage, and the stock price tends to deviate from its equilibrium value, fueling so-called speculative bubbles. The correction process that brings the asset price back to its equilibrium value is arbitrage: the competitive behavior of rational investors ensures the *absence of arbitrage* opportunities, that is, it eliminates the possibility of systematically making profits without incurring costs and/or risks (*money pump*). The resulting portfolio recomposition is reflected in the stock prices. The adjustment process stops when prices return to equilibrium values. At this point, investors no longer have an incentive to modify the composition of their portfolios.

3.4.1.1 The Limits of Efficiency: When Information is Costly

Over the years, several theoretical works have questioned whether the market can satisfy *strong* efficiency (Niederhoffer and Osborne 1966; Scholes 1969).[2] This criticism was advanced by Grossman and Stiglitz (1976, 1980), who studied the role of prices as a mechanism for market equilibrium and a process by which information is conveyed from informed to uninformed agents. This analysis does not differ in spirit from that of Akerlof (1970) on asymmetric information. The authors show that when information is costly, the hypothesis of strong efficiency fails. Grossman and Stiglitz's idea can be summarized as follows.

If information is costly, the marginal investor should be indifferent to being informed or uninformed. However, when no investors are informed, prices contain no information, and the value of information is very high. In contrast, when many agents are informed, prices contain extensive information; additional information thus has a very low value. Consequently, *in equilibrium* a share of agents acquires the information at a specific cost, a share decides not to bear additional costs, and at least one *marginal investor* is indifferent between the two options.

But, if markets are in equilibrium, how can investors who have incurred an additional cost earn arbitrage profits to compensate? "If the assumption of *strong* efficiency holds, prices reflect all the information available in the market. Therefore, incurring additional costs is suboptimal because the prices are so informative that new information is immediately incorporated. In this scenario, it is no longer con-

---

[2] To this end, less restrictive efficiency definitions were developed (Roberts 1959) and empirically tested (Fama 1970).

venient to pay for additional information. But at the same time, if no investor buys information, prices are not informative, and the financial market is inefficient and could even fail.

Therefore, Grossman and Stiglitz conclude that when information is *costly*, prices *do not* reflect all available information, and financial markets cannot operate with full efficiency.[3] However, the reduced efficiency should nevertheless allow investors to make profits to compensate for the additional costs of obtaining information. Thus, the portfolio allocation of informed agents will be better than that of uninformed ones. All in all, Grossman and Stiglitz modify the concept of market efficiency and clarify that the assumption of zero information cost is not only a *sufficient* condition for prices to reflect information but also a *necessary* condition to ensure the validity of the EMH.

3.5 Efficiency and Prices

We explore the implications of the EMH by starting from the intertemporal optimality condition of consumption when income and financial returns are uncertain. For convenience, we rewrite the corresponding Euler equation,

$$u'(c_t) = \frac{1}{1+\delta} E_t\left[Z_t u'(c_{t+1})\right], \tag{3.25}$$

where $Z_t = (1 + z_t)$ is the expected return of the risky asset and δ the intertemporal preference rate.

To explore this relation let us denote by p_t the price at time t of an asset measured in terms of consumption, and with d_t the dividend paid at the beginning of each period. d_t is a stochastic Markovian process that satisfies the property

$$F\left[d_{t+s+1} = j \mid d_t, \ldots, d_{t+s}\right] = F\left[d_{t+s+1} = j \mid d_{t+s}\right]. \tag{3.26}$$

This expression implies that the probability $F[\cdot]$ that the stochastic variable d_{t+s+1} is in state j, conditional on all past information $d_t, \ldots, d_{t+s}$, is equal to the probability that d_{t+s+1} is in state j, conditional only on information inferred from d_{t+s}. In other words, the Markovian property states that the present and future periods are independent: all that matters in determining the future value d_{t+1} is its present value d_t.

Now, we define the risky return as $Z_t = \frac{p_{t+1}+d_{t+1}}{p_t}$, where p_{t+1} and d_{t+1} are the values of the price and dividend in the subsequent period. Substituting Eq. 3.25 for Z_t we get

$$p_t u'(c_t) = \frac{1}{1+\delta} E_t\left[u'(c_{t+1})(p_{t+1} + d_{t+1})\right]. \tag{3.27}$$

[3] Grossman and Stiglitz point out that empirical tests of capital market efficiency show that there are no profit opportunities from time series analysis of stock prices. Some notable works that have dealt with the issue include Alexander (1961), Cootner (1964), Fama (1965) and Fama and Blume (1966).

In this form, the Euler equation states that the current price of the stock depends only on the expected values in $t+1$, once adjusted for dividends and weighted by marginal utilities. To simplify Eq. 3.27 further, we use the product decomposition of two normal random variables and assume that the utility function is linear. This implies that the ratio of intertemporal marginal utilities is constant over time and equal to $\frac{u'(c_{t+1})}{u'(c_t)} = 1$. With these assumptions, under the further hypothesis that $Cov\left[\frac{u'(c_{t+1})}{u'(c_t)}, \frac{p_{t+1}+d_{t+1}}{p_t}\right] = 0$, Eq. 3.27 becomes

$$p_t = \alpha E_t\left[p_{t+1} + d_{t+1}\right], \tag{3.28}$$

where $\alpha = \frac{1}{1+\delta}$ is the intertemporal discount factor. Note that the joint assumption of the Markovian process and risk neutrality allow us to derive a simplified evaluation formula. Equation 3.28 is a first-order finite difference equation with constant coefficients that can be solved iteratively by substitution. If the planning horizon reaches T, the solution is

$$p_t = \sum_{s=1}^{T} \alpha^s E_t\left(d_{t+s}\right) + \alpha^T E_t\left(p_{t+T}\right). \tag{3.29}$$

This expression states that a risk-neutral investor can purchase a dividend stream d_{t+s} plus the residual value of the stock at time T, paying (today) a price p_t. In other words, since, at the terminal time, the value of the stock p_{t+T} is equal to the discounted value of future dividends from that period *onward*, the value of dividends from the first periods plus the future *capital gain* is exactly equal to the discounted value at time t of the infinite dividend stream.

If the horizon is infinite, Eq. 3.29 can be further simplified because

$$\lim_{T\to\infty} \alpha^T E_t\left(p_{t+T}\right) = 0, \tag{3.30}$$

with $\alpha < 1$. This is the well-known *transversality* condition which eliminates the *speculative bubbles*. From Eq. 3.30 we then obtain

$$p_t = \sum_{s=1}^{\infty} \alpha^s E_t\left(d_{t+s}\right); \tag{3.31}$$

that is, the present value of the asset is equal to the discounted value of the infinite stream of dividends.

3.6 The Random Walk Theory of Prices

Suppose now that $\alpha \simeq 1$ and the stock has no associated dividends. In this case, the relation in Eq. 3.28 becomes

$$p_t = E_t \left[p_{t+1} \right], \tag{3.32}$$

or, equivalently, by removing the expectation operator

$$p_{t+1} = p_t + \varepsilon_{t+1}, \tag{3.33}$$

where the shock ε_{t+1} has a normal probability distribution with zero mean $E_t\left(\varepsilon_{t+1}\right) = 0$, constant variance, $Var_t\left(\varepsilon_{t+1}\right) = \sigma^2$, and is serially uncorrelated,

$$Cov \left[\varepsilon_t, \varepsilon_{t+1} \right] = 0. \tag{3.34}$$

In other words, we are assuming that the shocks follow the normal distribution $\varepsilon_t \backsim N\left(0, \sigma^2\right)$ (see Fig. 3.2). A stochastic dynamic with these properties is a Markovian process. Generally speaking, all stochastic models that satisfy this assumption fall into the class of processes referred to as *random walks*. In particular, all versions of the random walk hypothesis and martingale hypothesis are captured by Eq. 3.34, which can be seen as an orthogonality condition between the shocks.

3.6.1 The Martingale

Equation 3.32 captures the concept of *martingale*, a fair game in which no player has an advantageous position with respect to the outcome of the game. Formally, the

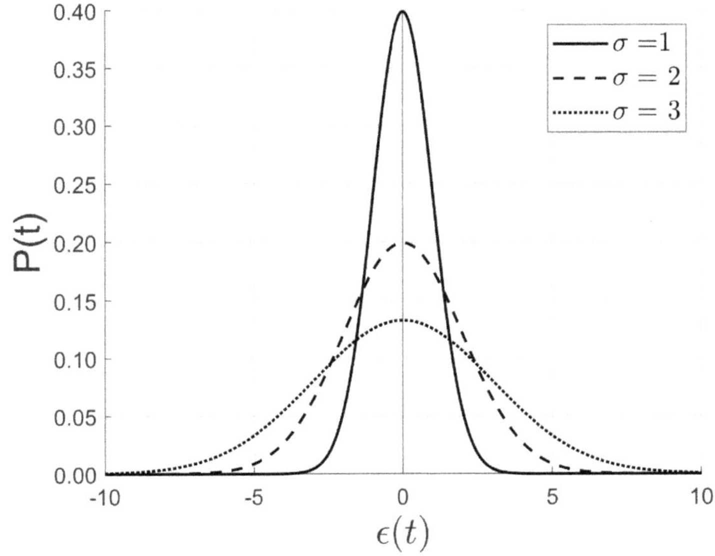

Fig. 3.2 The normal distribution of shock ε_t

fairness of a game is defined as a stochastic process p_t that satisfies the following condition

$$E_t\left[p_{t+1} \mid p_{0,\cdots,}p_t\right] = p_t \tag{3.35}$$

or

$$E_t\left[p_{t+1} - p_t \mid p_{0,\cdots,}p_t\right] = 0. \tag{3.36}$$

Equation 3.35, referred to as the *martingale property*, states that a game is fair if the game's expected value is, on average, equal to its current value, regardless of historical outcomes. Thus, fairness lies in the fact that the expected value of the price in the next period, $E_t\left[p_{t+1} \mid p_{0,\cdots,}p_t\right]$, is simply equal to the price p_t in the current period.

Similarly, Eq. 3.36 tells us that a game is a martingale if the expected value of the *change* in price is equal to zero. Therefore, the future price increases or decreases with the same probability. Clearly, these properties must be consistent with the functioning of the financial market. If the current price of a security is, for example, lower than expected, investors will find it worthwhile to buy that asset in the next period. However, the increase in demand causes a raise in the current price that will cease only when the price in t becomes equal to the expected price in $t + 1$.

Obviously, the mechanism just described implies that price is an information vehicle that conveys an asset's economic and financial characteristics to all investors. This property is coherent with the EMH.

3.6.2 Random Walk

Equation 3.33 describes a *random walk* dynamics. This is a special case of a martingale, under the assumption that the conditional variance of the process is constant over time. In its most general form, it is written as

$$p_t = a + p_{t-1} + \varepsilon_t, \tag{3.37}$$

where a is a deterministic growth rate and $\varepsilon_t \backsim N\left(0, \sigma^2\right)$ is the idiosyncratic shock (i.e., *white noise*) that occurs in each period. If the shocks, ε_t, are uncorrelated (condition in Eq. 3.24), subsequent price changes are independent and past history. Therefore, in each period t, the probability of the price going up or down is the same, and it is independent of the values in the previous periods.

Differentiating the process in Eq. 3.37 at the initial period ($t = 0$) yields the sequence

$$p_t - p_{t-1} = a + \varepsilon_t, \tag{3.38}$$

$$p_{t-1} - p_{t-2} = a + \varepsilon_{t-1}, \tag{3.39}$$

$$\cdots = \cdots$$

$$\text{and} \quad p_1 - p_0 = a + \varepsilon_1, \tag{3.40}$$

from which it follows that

$$p_t - p_0 = at + \sum_{i=1}^{t} \varepsilon_i. \tag{3.41}$$

Therefore, the mean and variance of the price, conditional on its initial value p_0, turn out to be

$$E\left[p_t \mid p_0\right] = at \tag{3.42}$$

$$\text{and} \quad Var\left[p_t \mid p_0\right] = \sigma^2 t \tag{3.43}$$

These two equations together show that the *random walk* is a stochastic *non-stationary* process since both mean and variance are functions of time. Observe that if $a = 0$, the process is stationary in the mean, but the variance remains a linear function of time.

The importance of martingales and random walk processes does not lie in their normative properties but rather in their ability to capture stylized facts. In a famous article, (Kendall and Hill 1953, p. 13) demonstrates that the behavior of stock and security prices do not indicate the existence of regular cycles and do not offer the possibility of defining algorithms that are useful for predicting future price levels. Indeed, each set of prices looks like "a 'wandering' one, almost as if once a week the Demon of Chance drew a random number [...] and added it to the current price to determine the next week's price". In other words, prices seemed to follow a random walk dynamic.[4]

At that time, this conclusion was one that surprised most scholars and practitioners. This result can be explained as follows. If prices capture all relevant information available in each period, then their variation reflects that new information has been introduced. However, by definition, this new information cannot be predicted (since it would then not be new), and consequently, price changes cannot be anticipated. Therefore, if the current price of a security incorporates all information, price changes in that security must necessarily reflect only information that could not have been predicted. This, in turn, means that the stochastic process governing price realization must be a random walk.

Hence, when we refer to the EMH, we are referring to precisely what we have just stated above. A market is efficient when the prices of securities fully and correctly reflect all available information at all times. In such a financial market, neither *fundamental analysis*—which studies trends in the value of firms by analyzing their profitability to find out whether there are new prospects for their stock value—nor *technical analysis*—which studies past prices with the aim of predicting future prices—allow an investor to make greater profits than another investor would with a

[4] It should be noted that the idea that stock prices follow a random path was first put forward by Bachelier (1900) in his doctoral dissertation. Later, Holbrook Working (1934) observed a similar pattern for commodities.

portfolio of randomly chosen securities with the same degree of probability. This is why the idea of an efficient financial market is associated with the idea of the *random walk*.

At this point, it is worth noting that some market conditions may hinder or facilitate price adjustment in the presence of new information. The following conditions, if met, are *sufficient* to ensure the existence of an efficient market: (1) absence of *transaction costs*, (2) *zero information cost* for all agents in the market, (3) *homogeneity of expectations* regarding the effect of available information on current and future prices of securities.

The purpose of the empirical analysis is to understand whether (and to what extent) the absence of such conditions can shift the market away from equilibrium and efficiency.

3.7 Empirical Estimations

In this next section, we maintain the assumption of efficient markets. However, we focus on the empirical implications of the CCAPM by rewriting the condition in Eq. 3.11 as follows:

$$1 = E_t\left[Z_t \frac{u'(c_{t+1})}{(1+\delta)\,u'(c_t)}\right]; \tag{3.44}$$

that is

$$1 = E_t\left(Z_t m_{t+1}\right), \tag{3.45}$$

where the term $\frac{u'(c_{t+1})}{(1+\delta)u'(c_t)} \equiv m_{t+1}$ measures the marginal rate of intertemporal substitution, also known as the *stochastic discount factor*. In other words, the component m_{t+1} discounts at the present time the uncertain return Z_t that is obtained from an asset with price 1 at time t. The factor m_{t+1} generalizes the discount term: risk adjustment depends on the correlation between the stochastic components of consumption and returns. Note, that for each utility function, there will be a different stochastic discount factor and, thus, a different function relating m_{t+1} to the observed data. To evaluate the empirical robustness of Eq. 3.44, it is necessary to analyze some alternative utility functions.

3.7.1 The Constant Relative Risk Aversion Function

We begin with the constant relative risk aversion (CRRA) function:

$$u(c_t) = \frac{c_t^{1-\gamma}}{1-\gamma}, \tag{3.46}$$

where, as usual, γ is the coefficient of relative risk aversion (Mehra and Prescott 1985). Substituting for $u'(c_{t+s})$ the previous condition becomes

$$1 = \frac{1}{1+\delta} E_t \left[Z_t \left(\frac{c_{t+1}}{c_t} \right)^{-\gamma} \right], \tag{3.47}$$

which, expressed in logarithms and using the approximation $\ln(1+x) \simeq x$, reduces to

$$0 = -\delta + \ln E_t \left[Z_t \left(\frac{c_{t+1}}{c_t} \right)^{-\gamma} \right]. \tag{3.48}$$

We simplify Eq. 3.48 further by assuming that the joint probability distribution of returns and stochastic discount is lognormal and homoscedastic. Under the lognormality assumption, the expected value of any stochastic variable x can be approximated using Taylor's theorem:

$$\ln E_t(x) = E_t \ln(x) + \frac{1}{2} Var_t \ln(x), \tag{3.49}$$

where $Var_t \ln(x) = E_t \left[(\ln(x) - E_t(\ln(x)))^2 \right]$. In our case of two lognormal variables, as in Eq. 3.48, the corresponding transformation is

$$0 = -\delta + E_t \ln(Z_{it}) - \gamma E_t(g) + \frac{1}{2} Var_t \ln \left[Z_t \left(\frac{c_{t+1}}{c_t} \right)^{-\gamma} \right], \tag{3.50}$$

where $g = \Delta \ln c_{t+1}$ denotes the growth rate of consumption between t and $t+1$. The overall variance is given by

$$Var_t \ln \left[Z_{it} \left(\frac{c_{t+1}}{c_t} \right)^{-\gamma} \right] = \sigma_i^2 + \gamma^2 \sigma_c^2 - 2\gamma\sigma_{ic}, \tag{3.51}$$

where σ_i^2 is the variance of the return of the i-th risky security, σ_c^2 is the variance of consumption, and σ_{ic} is the covariance of the two stochastic variables.

3.7.2 The Risky Asset

For the risky security i, it can be written that

$$E_t(z_{it}) = \delta + \gamma E_t(g) - \frac{1}{2} \left(\sigma_i^2 + \gamma^2 \sigma_c^2 - 2\gamma\sigma_{ic} \right). \tag{3.52}$$

This relation states that, for the CCAPM, the expected growth rate of consumption $\gamma E_t(g)$ affects the value of all assets. The difference in returns is then measured by the variance of z_{it} and the covariance between returns and consumption.

When the utility function is a CRRA, the variance of returns (σ_i^2) and consumption (σ_c^2) decreases the equilibrium value of expected return $E_t(z_{it})$; but the covariance σ_{ic} can increase or decrease the expected return $E_t(z_{it})$ depending on whether it is positive or negative.

If σ_{ic} is *positive*, the asset i yields more in periods when consumption is increasing, that is, when $u'(c_t)$ decreases, thus *destabilizing* consumption. This characteristic of asset i must, therefore, be offset by a higher expected return for risk-averse individuals to be willing to hold it in their portfolio. For the opposite reason, a *negative* covariance, *stabilizes* consumption over time and requires a lower risk premium.

3.7.3 The Risk-Free Asset

Equation 3.52 can also be used to identify paradoxes of *asset pricing*. Consider, for instance, the risk-free asset that offers a return equal to $R_t = (1 + r_t)$. For this security, the variance σ_i^2 and the covariance σ_{ic} are obviously zero, and the non-risky interest rate meets the relation

$$r_t = \delta + \gamma E_t(g) - \frac{1}{2}\gamma^2\sigma_c^2. \tag{3.53}$$

The value of r_t depends on the intertemporal preference rate δ, the consumption growth rate and its variability, σ_c^2. More precisely, given δ, a higher rate of consumption growth requires a higher rate r_t to offset agents' incentives to transfer resources to the present, limiting the difference between current and future levels of consumption. Note that the term $-\frac{1}{2}\gamma^2\sigma_c^2$ measures the weight of consumption variance. For this reason, it is usually interpreted as a parameter related to the *precautionary component* of savings (see Chap. 2, Sect. 2.5).

3.7.4 Equity Premium Puzzle

Subtracting Eq. 3.53 from Eq. 3.52 yields the *excess* of the risky return over the riskless return

$$[E_t(z_{it}) - r_t] + \frac{\sigma_i^2}{2} = \gamma\sigma_{ic}. \tag{3.54}$$

The variance $\frac{\sigma_i^2}{2}$ that appears on the left-hand side of Eq. 3.54 depends solely on the assumption of lognormality in the distribution of returns. Clearly, the value of the expectation depends on the risk aversion coefficient, γ.

This equation was used by Mehra and Prescott (1985) to illustrate the so-called *equity premium puzzle* (EPP). Specifically, these authors show that for the observed values of σ_i^2 and σ_{ic} of U.S. equities and consumer stocks, relative to risk-free stocks, Eq. 3.54 is satisfied only for values of γ *so high* as to be incompatible with a plausible consumer attitude toward risk.

Table 3.1 Aversion coefficient estimates

Country	Sample period	γ
France	1973Q2–2011Q2	<0
Germany	1978Q4–2011Q2	<0
Italy	1971Q2–2011Q2	66.96
Japan	1970Q2–2011Q2	1183.09
Netherlands	1977Q2–2011Q2	141.29
Sweden	1970Q1–2011Q2	314.53
UK	1970Q1–2011Q2	<0
US	1947Q2–2011Q2	154.98

Note γ is the relative risk aversion coefficient. This is obtained by dividing the average adjusted excess return by the estimated covariance
Source Data taken from Campbell (2017)

This anomaly was subsequently confirmed by Mankiw and Zeldes (1991), using data from the U.S. financial market for the period 1890–1979. Specifically, their calculations show that the expected excess return is $[E_t(z_{it}) - r_t] = 0.06$, while the mean square deviation of risky returns is $\sigma_i = 0.036$, and that of consumption $\sigma_c = 0.167$. Finally, the index of correlation between the two variables is $\rho_{ic} = 0.4$. Based on these values, the covariance between yield and consumption is

$$\sigma_{ic} = \rho_{ic}\sigma_i\sigma_c = 0.0024. \tag{3.55}$$

Equation 3.55 implies that to meet the *asset pricing* relationship derived from the CCAPM, the degree of relative risk aversion should be equal to

$$0.06 = \gamma 0.0024 - \frac{0.00129}{2}; \tag{3.56}$$

that is, $\gamma = 25.27$. This estimate is obviously problematic since it generates paradoxical results about individuals' risk aversion. For example, for $\gamma = 25$, it turns out that an individual prefers a 17% reduction in consumption to the alternative of seeing his or her consumption be reduced by 20% with a probability of 0.5.

Importantly, this anomaly becomes even more pronounced when the empirical analysis is developed using economic and financial data for the past 50–60 years (Table 3.1). Several empirical studies have also confirmed the previous paradox using data from other countries. For all of these studies, it appears that γ is far above the value of 10, the maximum level of risk aversion considered plausible by Mehra and Prescott (1985). This fact has become famous in the literature as the EPP.

3.7.5 Risk-Free Rate Puzzle

Plausible values of γ can be derived from Eq. 3.54 by assuming that r_t is extremely high. However, the latter assumption is contradicted by historical data (Table 3.2).

Table 3.2 Time preference rates

Country	Sample period	r_t	δ
France	1973Q2–2011Q2	2.08	−17.80
Germany	1978Q4–2011Q2	2.38	−5.97
Italy	1971Q2–2011Q2	1.86	−8.73
Japan	1970Q2–2011Q2	1.03	−8.15
Netherlands	1977Q2–2011Q2	2.29	−8.86
Sweden	1970Q1–2011Q2	1.68	−17.84
UK	1970Q1–2011Q2	1.34	−20.48
US	1947Q2–2011Q2	0.72	−37.97

Note r_t is the short-term log interest rate and δ is the time preference rate
Source Data taken from Campbell (2017)

Further, if γ is very high, there is a *second paradox* involving the Eq. 3.53. To illustrate this anomaly, we omit, for the moment, the term $-\frac{1}{2}\gamma^2\sigma_c^2$ and rewrite Eq. 3.53 as

$$r_t = \delta + \gamma E_t (g) . \tag{3.57}$$

Keep in mind that the consumption time series is characterized by a positive *drift*. Consequently, a high value of γ implies a high value of $\gamma E_t (g)$. Equation 3.57 shows that, to reconcile the historical data of r_t with the estimated data, we must assume that the intertemporal preference rate δ is *negative* such that the difference between δ and $\gamma E_t (g)$ generates a low value of r_t, consistent with that observed data (Table 3.2).

The anomaly of this result lies in the fact that, for a risk-averse agent with a CRRA utility function, we should observe a preference toward saving and, thus, a resistance to anticipate consumption over time (that is, a resistance to taking on additional debt). This would require that $\delta > r$ and that the time profile of consumption is decreasing. However, the existence of a *positive* historical average consumption growth rate, a *negative* intertemporal preference rate δ and a *positive* short-term interest rate r_t contradict this implication. This contradiction between historical fact and theoretical implication is referred to Weil (1989) as the *risk-free rate puzzle* (RRP).

Finally, it is important to note that the presence of the negative term $-\frac{1}{2}\gamma^2\sigma_c^2$ does not change the previous result. In fact, the quadratic term reflects the precautionary component of saving: risk-averse individuals, in the presence of uncertain consumption, prefer to save beyond the amount that guarantees *consumption smoothing*. This extra component of saving further reduces the incentive to anticipate future income to the present.

3.8 Alternative Hypotheses on Preferences

Researchers have attempted to solve the EPP and RRP puzzles by adopting alternative assumptions about the shape of the utility function. In particular, considerable effort has gone into eliminating the assumption of *additivity* and *separability* of preferences. These contributions can be divided into three main groups:

1. those in which a generalized expected utility function is used to *separate* the degree of risk aversion and elasticity of intertemporal substitution (EIS);
2. those in which a functional form of preferences is employed to incorporate *consumption habits*;
3. those in which the *expected utility* model is replaced by *behavioral* models of *consumption and saving*.

3.8.1 Generalized Expected Utility

The analysis in the previous section shows that the CRRA utility function used by Mehra and Prescott (1985) is consistent with the observed data *only if* the degree of relative risk aversion is excessively high. As we saw in Chap. 1, a strong restriction imposed by this class of utility functions is that the EIS is equal to the reciprocal of relative risk aversion.

The main implication of this property is that if, in the presence of uncertainty, an individual is reluctant to change his or her consumption decision in a period (due to a high γ), then is also reluctant to change his or her consumption over time (due to a low $1/\gamma$). Clearly, there is no reason why this should be a plausible hypothesis. Risk aversion describes a consumer's reluctance to substitute consumption in different states of the world and is a suitable measure to describe choices under uncertainty, even in a static model. EIS, by contrast, describes the willingness of consumers to substitute consumption *over time* and is also a useful index in a deterministic model. For this reason, some authors have tried to break the link between the two parameters.

Epstein and Zin (1989, 1991) propose the adoption of a *generalized expected utility* (GEU) function, defined as

$$u_t = \left\{ (1-\beta)\, c_t^{\frac{1-\gamma}{\theta}} + \beta \left(E_t u_{t+1}^{1-\gamma} \right)^{\frac{1}{\theta}} \right\}^{\frac{\theta}{1-\gamma}}, \tag{3.58}$$

where $\beta = \frac{1}{1+\delta}$, $\theta = (1-\gamma)\,/\,(1-1/\psi)$, γ is the degree of relative risk aversion, ψ is the EIS, and δ is the intertemporal preference rate. In this function, the current utility level is a function with constant elasticity between current consumption and future utility. We note that when $\gamma = 1/\psi$, then $\theta = 1$ and Eq. 3.58 becomes linear. The forward resolution of Eq. 3.58 reduces to the standard model with an additive and separable utility function. The main advantage of using the generalized form is that a high value of γ does not necessarily imply that individuals find it optimal to stabilize consumption over time. This property can solve the EPP puzzle. We consider this in detail.

The budget constraint takes the form

$$a_{t+1} = (1 + r_{mt})\,(a_t - c_t)\,, \tag{3.59}$$

where r_{mt} denotes the total return from the m portfolio. The corresponding Euler equation is

$$1 = E_t \left\{ \left[\beta \left(\frac{c_{t+1}}{c_t} \right)^{\frac{1}{\psi}} \right]^{\theta} \left[\frac{1}{(1 + r_{mt})} \right]^{1-\theta} (1 + z_{it}) \right\}, \tag{3.60}$$

where $(1 + z_{it})$ is the risky return of the i-th stock. From this equation, under the assumption of lognormality of returns and consumption, we determine that the non-risky interest rate must satisfy the relationship

$$r_t = \delta + \frac{1}{\psi} E_t (g) - \left(\frac{\theta}{2\psi^2} \sigma_c^2 - \frac{\theta - 1}{2} \sigma_m^2 \right). \tag{3.61}$$

Therefore, the value of the risk-free security return is given by δ, plus $\frac{1}{\psi}$ for the growth rate of consumption minus a constant that depends on the variances. Note that for $\frac{1}{\psi} = \gamma$ and $\theta = 1$, we find the relation in Eq. 3.53 derived from the CRRA function.

In this context, the corresponding appropriate *risk premium* turns out to be

$$[E_t (z_{it}) - r_t] + \frac{\sigma_i^2}{2} = \theta \frac{\sigma_{ic}}{\psi} + (1 - \theta) \sigma_{im}, \tag{3.62}$$

that is, the risk premium of security i is a weighted mix of the two covariances of the security with consumption (divided by EIS) and market portfolio m. The weights are θ and $(1 - \theta)$, respectively.

Equations 3.61 and 3.62 seem to offer a solution to the two paradoxes discussed above. Regarding the RRP in Eq. 3.61, a high relative risk aversion has *no implication* for the average level of non-risky interest. This is because, with the recursive function, the expected growth rate of consumption $E_t (g)$ is divided by ψ, which is independent of the value of γ.

In relation to the EPP, note instead that Eq. 3.62 does not require the existence of a high level of risk aversion to explain the risk premium. In fact, if $\theta \neq 1$, the risk premium depends on the value of the two covariances, σ_{ic} and σ_{im}, and the EIS. Moreover, the observed variability of wealth is much higher than that of consumption—the value of σ_{im} is greater than σ_{ic}—and this can empirically explain the existence of the risk premium.

Unfortunately, the relevance of these results is limited by some notable difficulties, of which the main three are the following:

1. First, as far as EPP is concerned, empirical analyses show that the value of ψ is low and θ is high. This implies that to estimate a reasonable risk premium, empirical verification should attach great importance to the covariance between consumption and returns, while stylized facts show that σ_{ic} is rather small.

2. This last stylized fact creates a further problem. Keep in mind that consumption and wealth are linked by the budget constraint; if consumption tends to be flat while wealth is volatile, a further paradox must be explained. Indeed, the latter cannot be taken as an exogenous fact to explain EPP.
3. Finally, note that in testing this alternative preference structure, considerable importance must be attached to the expected value of utility in $t+1$. This makes the calibration of the present model extremely difficult to implement. For example, Mehra (2003) argues that the proxies used by Epstein and Zin to calibrate the generalized expected utility of the later period are overestimated.

3.8.2 Consumption Habits

A second approach is to change the preference structure by incorporating consumption habits into the utility function (Constantinides 1990; Gali 1994; Heaton 1995; Alessie and Lusardi 1997). In this case, while still adopting a CRRA-type function, the assumption of temporal separability of preferences is removed, and utility is assumed to be influenced by both current and past consumption.

A typical form of these preference functions is

$$u(c_t) = E_t \sum_{s=0}^{\infty} \frac{(c_{t+s} - \phi c_{t+s-1})^{1-\gamma}}{1-\gamma} \qquad \text{con } \phi > 0, \tag{3.63}$$

where ϕ is a parameter representing the effects of *past consumption* on current consumption. This formalization captures the idea that utility is a decreasing function of past consumption, while current marginal utility is an increasing function of consumption in previous periods. This formulation also solves the conundrum of the *risk-free rate* since the additional degree of risk aversion implies an increase in the demand for non-risky securities and, thus, a reduction in the certain interest rate.

The simplest way to write a utility function that depends on consumption habits is by assuming that each consumer has a baseline level of consumption that is defined by the average level of consumption in the relevant social group. With this assumption, the consumption utility in each period becomes a *relative* function of the gap between individual consumption and that of other individuals belonging to the same class (desired level). The advantage of using this type of function is that individuals are especially resistant to changes in consumption relative to the reference, even when the relative risk aversion γ is very low. Abel (1990) defines this approach as "keeping up with the Joneses".[5] This reference level can be referred to as the *norm* (or even fashion) of consumption.

[5] The sentence originates from the comic strip "Keeping Up with the Joneses", created by Arthur R. Momand in 1913. The strip depicts the McGinis family, who are struggling to "keep up" with their neighbors, the Joneses of the title.

In the latter case, the utility function can be written as

$$u(c_t) = \frac{(c_t - N_t)^{1-\gamma}}{1-\gamma}, \tag{3.64}$$

where N_t is the norm of consumption. If we define the magnitude $n_t = (c_t - N_t)/c_t$ as a percentage deviation from N_t, the utility function in Eq. 3.64 becomes

$$u(c_t, N_t) = \frac{(n_t c_t)^{1-\gamma}}{1-\gamma}, \tag{3.65}$$

with the first derivative with respect to c_t equal to $u_{c_t} = (c_t - N_t)^{-\gamma} = (n_t c_t)^{-\gamma}$. The first-order condition is

$$1 = \frac{1}{1+\delta} E_t \left[Z_{it} \left(\frac{n_{t+1}}{n_t} \right)^{-\gamma} \left(\frac{c_{t+1}}{c_t} \right)^{-\gamma} \right], \tag{3.66}$$

which holds for each security i. Let us now assume that the time evolution of the norm is given by the expression

$$\Delta \lg (n_{t+1}) \equiv \eta_t = \alpha \varepsilon_{t+1}. \tag{3.67}$$

Consumption follows the dynamic

$$\Delta \lg (c_{t+1}) \equiv g_t = \pi + \varepsilon_{t+1} \tag{3.68}$$

where ε_t is an idiosyncratic shock characterized by the property $\varepsilon_t \backsim N(0, \sigma^2)$ and α is the parameter that captures the effects of consumption habits. If n_t and c_t have a lognormal distribution, then Eq. 3.66 becomes

$$\begin{aligned} 0 = & -\delta + E_t(z_t) - \gamma E_t(\eta_t) - \gamma E_t(g_t) \\ & + \frac{1}{2} Var_t \lg \left[Z_{it} \left(\frac{n_{t+1}}{n_t} \right)^{-\gamma} \left(\frac{c_{t+1}}{c_t} \right)^{-\gamma} \right]. \end{aligned} \tag{3.69}$$

For the three lognormal variables, the overall variance is given by the following expression:

$$Var_t \ln(\cdot) = \sigma_i^2 + \gamma^2 \sigma_c^2 + \gamma^2 \sigma_n^2 - 2\gamma \sigma_{ic} - 2\gamma \sigma_{in} - 2\gamma^2 \sigma_{cn}. \tag{3.70}$$

Equation 3.70 can then be rewritten as

$$E_t(z_t) = \delta + \gamma \pi - \frac{1}{2} \left(\sigma_i^2 + \gamma^2 \sigma_c^2 + \gamma^2 \sigma_n^2 \right) + \left(\gamma \sigma_{ic} + \gamma \sigma_{in} + \gamma^2 \sigma_{cn} \right). \tag{3.71}$$

For the risk-free asset, this reduces to

$$r_t = \log\delta + \gamma\pi - \frac{1}{2}\gamma^2\left(\sigma_c^2 + \sigma_n^2 - 2\sigma_{cn}\right). \tag{3.72}$$

Finally, subtracting Eq. 3.71 from Eq. 3.72 yields an expression for the *risk premium*, which is influenced by consumption habits:

$$\left[E_t(z_t) - r_t\right] + \frac{\sigma_i^2}{2} = \gamma\left(\sigma_{ic} + \sigma_{in}\right); \tag{3.73}$$

that is, given that $\sigma_{in} = \alpha\sigma_{ic}$,

$$\left[E_t(z_t) - r_t\right] + \frac{\sigma_i^2}{2} = \gamma(1+\alpha)\sigma_{ic}. \tag{3.74}$$

This result can resolve the EPP anomaly. Indeed, consumption habits, captured by α, can explain the risk premium without the need for implausible assumptions about the value of γ. If, for example, risky securities have a negative return, they cause a reduction in individual consumption (σ_{ic} is positive) and a deviation from the norm, greatly reducing utility. Therefore, the higher the individual's sensitivity to his or her consumption deviating from the norm, the higher the risk premium they require to hold risky securities in the portfolio. Simulations using this model (see Campbell 1998) have shown that plausible parameter values tend to reabsorb the empirical paradoxes associated with EPP.

Finally, regarding the RRP, note that substituting Eq. 3.68 into Eq. 3.69 yields the relationship:

$$\eta_t = \alpha\left[g_t - \pi\right]. \tag{3.75}$$

This latter expression states that consumption habit measured by η_t changes only if the growth rate of consumption g_t moves away from the deterministic value of growth, π. Hence, a high sensitivity of the norm (i.e., a high value of α) to innovations in c_t increases the importance of the precautionary component of saving, as measured by σ_c^2. Equation 3.72, related to the risk-free interest rate r_t captures this implication. A higher level of precautionary saving results in an increasing demand for securities with a lower expected return (reduction measured by the value of the component $-\frac{1}{2} - \sigma_c^2$).

3.8.3 Behavioral Hypotheses

A third way to resolve the CCAPM paradoxes is to modify the analytical structure under uncertainty by assuming that agents give *more weight* to negative than to positive events. For example, faced with a choice between different portfolios, a risk-averse individual would be more influenced in his or her decision by the possibility of incurring losses than by the prospect of large returns. Is this an acceptable assumption? Empirical analysis shows that the composition of household financial wealth is systematically characterized by a small proportion of risky assets compared to risk-free assets, even though the return on risky assets, such as stocks, is, on average, higher than the *risk-free* interest rate.

Why do investors hold only a limited share of stocks in their portfolios? We know that CCAPM fails to answer these paradoxes and that the interpretation of empirical "anomalies" requires a high and implausible level of risk aversion.

Obviously, the trend of excess returns over time is characterized by volatility. For example, Dimson et al. (2002) showed that the standard deviation of the risk premium, estimated from the difference $E_t(z_t) - r_t$, is relatively high in Italy (32%) and Germany (35%) but lower in the U.S., France, and the U.K. (about 20%). Thus, some evidence suggests that the undersized share of risky assets in financial portfolios also depends on individual perceptions of the combination of expected returns and riskiness (Jordá et al. 2019).

Behavioral *finance* emerged in the last 50 years in response to the empirical difficulties encountered by decision theory in considering the case of uncertainty. This new approach was intended to challenge the standard analysis described by the theory of expected utility. In the field of financial theory, Benartzi and Thaler (1995) succeeded in reconciling theory and data by assuming that agents exhibit relevant "loss aversion" based on *prospect theory* (Kahneman and Tversky 1979). More recently, Gul (1991) took a rigorous approach, studying portfolio allocation when household preferences are characterized by *disappointment risk aversion* (DRA). The results are interesting as the simulations produce *reasonable* portfolio allocations even for moderate values of risk aversion.

The intuition here is simple. Under the DRA hypothesis, the agent gives more weight to losses and is, therefore, less likely to invest in risky assets. Compared with previous behavioral theories, DRA has the *advantage* of being an axiomatic theory, the most interesting implication of which is that it provides endogenous threshold values for individual portfolio decisions. However, the *disadvantage* lies in the fact that endogeneity itself only rarely provides closed-form solutions. Therefore, numerical analyses are the standard tool for studying portfolio choices with DRA preferences. It is possible, however, to provide closed-form solutions even with DRA by studying the case of "small" risks (Saltari and Travaglini 2009) that concretely characterize the portfolio choices of individual (consumer) risk takers.

To explain DRA, we study a simple case of portfolio choice in which returns follow a stochastic binomality.

3.8.4 DRA and Portfolio Choice

We use a model with two states and pay-offs x_1 and x_2. We define the *DRA* expected utility $V(\alpha)$ as

$$V(\alpha) = p_1 u(w_0 + \alpha x_1) + p_2 u(w_0 + \alpha x_2) - \beta p_2 [V(\alpha) - u(w_0 + \alpha x_2)], \tag{3.76}$$

where $x_1 > 0 > x_2$. The last term in Eq. 3.76 captures the effect of disappointment. When the "bad" outcome, x_2, occurs, the agent is disappointed; his or her expected utility is reduced by a term that is the product of the disappointment aversion β and the expected disappointment. Note that when $\beta = 0$, this expression reduces to the traditional expected utility; w_0 is the initial level of wealth.

Making explicit the expression for $V(\alpha)$, yields

$$V(\alpha) = p_1 \frac{1}{1 + p_2\beta} u(w_0 + \alpha x_1) + p_2 \frac{1 + \beta}{1 + p_2\beta} u(w_0 + \alpha x_2). \tag{3.77}$$

Note that DRA implies that $V(\alpha)$ depends on the modified value of the original probabilities p_1 and p_2. These *new* probabilities, $p_1 \frac{1}{1+p_2\beta}$ and $p_2 \frac{1+\beta}{1+p_2\beta}$, attach more weight to the unfavorable event and less weight to the favorable event. It is useful to rewrite the previous expression as

$$V(\alpha) = q_1 u(w_0 + \alpha x_1) + q_2 u(w_0 + \alpha x_2), \tag{3.78}$$

where

$$q_1 = p_1 \frac{1}{1 + p_2\beta} \quad \text{and} \quad q_2 = 1 - q_1. \tag{3.79}$$

We refer to q_1 and q_2 as *disappointing* probabilities. The first-order condition for portfolio choice with DRA expected utility is

$$V'(\alpha_D^*) = q_1 x_1 u'(w_0 + \alpha_D^* x_1) + q_2 x_2 u'(w_0 + \alpha_D^* x_2) = 0, \tag{3.80}$$

where α_D^* is the optimal portfolio choice. We obtain the solution by computing the first-order condition when $\alpha_D^* = 0$:

$$\begin{aligned} V'(0) &= q_1 x_1 u'(w_0) + q_2 x_2 u'(w_0) \\ &= u'(w_0) E_D(x) = 0, \end{aligned} \tag{3.81}$$

where $E_D(x)$ is the *disappointing* expected value of x, computed using the modified probabilities. Since the marginal utility of wealth is positive, the first-order condition is satisfied if the adjusted expected value of x, $E_D(x)$, is equal to zero. Using the definitions in Eq. 3.79, this implies that

$$\begin{aligned} E_D(x) &\equiv q_1 x_1 + q_2 x_2 \\ &= \frac{E(x) + p_2 x_2 \beta}{1 + p_2\beta} = 0 \quad = u'(w_0) E_D(x) = 0. \end{aligned} \tag{3.82}$$

Thus, in order to have $\alpha_D^* = 0$, it must be

$$E(x) = -p_2 x_2 \beta. \tag{3.83}$$

In contrast to the traditional portfolio model based on expected utility theory, the amount of wealth invested in the risky asset is now equal to zero *if, and only if,* the expected value $E(x)$ of the excess return is equal to the expected disappointment, $-p_2x_2 > 0$, times the disappointment aversion, β. Under DRA, it might then be better not to invest in the risky asset even when its expected return is positive.

Further Reading

Among finance textbooks with extensive introductory material on portfolio choices, we recommend *Modern Portfolio Theory and Investment Analysis* by Elton et al. (2013) (Chaps. 1–3) and Investments by Sharpe and Alexander (1999) (Chaps. 1–3). On expected utility, risk aversion, and the standard portfolio model, see Gollier (2004) (Chaps. 1–4). Personal preferences in uncertain markets, the mean-variance model, and the Consumption CAPM are well discussed in Peleg (2014) (Chaps. 11–13), while informative notes on the recent development of the CAPM can be found in the article by Perold (2004). For a detailed but accessible exposition, which includes several extensions of the model, see instead Fama (1976) and Jagannathan and McGrattan (1995) as a good guide to empirical work. Notably, Fama (1993, 2004) and Fama and French (1992) stimulated much of the, still ongoing, applied research in the field, with periodic issues on the *Journal of Finance* and the *Journal of Financial Economics*. A noteworthy contribution in this area is Mehra (2003), which offers a critical examination of the EPP and its solution. Finally, a textbook that covers the entire spectrum of empirical finance is Campbell et al. (1997).

References

Abel, A. B. (1990). Asset prices under habit formation and catching up with the joneses. *National Bureau of Economic Research - Working Paper*, (3279).

Akerlof, G. A. (1970). The market for "lemons": Quality uncertainty and the market mechanism. *The Quarterly Journal of Economics, 84* (3), 488–500.

Alessie, R., & Lusardi, A. (1997). Consumption, saving and habit formation. *Economics letters, 55* (1), 103–108.

Alexander, S. (1961). Price movements in speculative markets: Trends or random walks. *Industrial Management Review, 2* (2), 7–26.

Arrow, K. J. (1996). The theory of risk-bearing: Small and great risks. *Journal of risk and uncertainty, 12* (2/3), 103–111.

Bachelier, L. (1900). *Théorie de la spéculation*. Self-published (PhD thesis).

Benartzi, S., & Thaler, R. H. (1995). Myopic loss aversion and the equity premium puzzle. *The Quarterly Journal of Economics, 110* (1), 73–92.

Campbell, J. Y., Lo, A. W. & MacKinlay, A. C. (1997). *The econometrics of financial markets*. Princeton University Press.

Campbell, J. Y. (2017). *Financial decisions and markets: A course in asset pricing*. Princeton University Press.

Campbell, J. Y., & Cochrane, J. H. (1998). By force of habit: A consumption- based explanation of plantation of aggregate stock market behavior. *Center for Research in Security Prices (CRSP) Working Paper*, (412).

Constantinides, G. (1990). Habit formation: A resolution of the equity premium puzzle. *Journal of Political Economy, 98* (5), 519–543.

Cootner, P. H. (1964). *The random character of stock market prices*. MIT University Press.
Dimson, E., Marsh, P., & Staunton, M. (2002). *Triumph of the optimists: 101 years of global investment returns*. Princeton University Press.
Elton, E. J., Gruber, M. J., Brown, S. J., & Goetzmann, W. N. (2013). *Modern portfolio theory and investment analysis. Ninth edition*. Wiley.
Epstein, L. G., & Zin, S. E. (1989). Substitution, risk aversion, and the temporal behaviour of consumption and asset returns: A theoretical framework. *Econometrica, 57* (4), 937–969.
Epstein, L. G., & Zin, S. E. (1991). Substitution, risk aversion, and the temporal behavior of consumption and asset returns: An empirical analysis. *Journal of Political Economy, 99*, 263–286.
Fama, E. (1965). The behavior of stock market prices. *Journal of Business, 38* (1), 34–105.
Fama, E. (1970). Efficient capital markets: A review of theory and empirical work. *Journal of Finance, 25*, 383–417.
Fama, E. (1976). *Foundations of finance*. Oxford: Basil Blackwell.
Fama, E. (1993). Common risk factors in the returns on stocks and bonds. *Journal of Financial Economics, 33* (1), 3–56.
Fama, E. (2004). The capital asset pricing model: Theory and evidence. *Journal of Economic Perspectives, 18* (3), 25–46.
Fama, E., & Blume, M. E. (1966). Filter rules and stock-market trading. *The Journal of Business, 39* (1), 226–241.
Fama, E., & French, K. R. (1992). The cross-section of expected stock returns. *Journal of Finance, 47* (2), 427–465.
Gali, J. (1994). Keeping up with the joneses: Consumption externalities, portfolio choice, and asset prices. *Journal of Money, Credit and Banking, 26* (1), 1–8.
Gollier, C. (2004). *The economics of risk and time*. MIT University Press.
Grossman, S. J., & Stiglitz, J. E. (1976). Information and competitive price systems. *The American Economic Review, 66* (2), 246–253.
Grossman, S. J., & Stiglitz, J. E. (1980). On the impossibility of informationally efficient markets. *The American Economic Review, 70* (3), 393–408.
Gul, F. (1991). A theory of disappointment aversion. *Econometrica*, 667–686.
Heaton, J. (1995). An empirical investigation of asset pricing with temporally dependent preference specifications. *Econometrica, 63* (4), 681–717.
Jagannathan, R., & McGrattan, E. R. (1995). The capm debate. *Federal Reserve Bank of Minneapolis Quarterly Review, 19* (4), 2–17.
Jensen, M. C. (1978). Some anomalous evidence regarding market efficiency. *Journal of Financial Economics, 6* (2/3), 95–101.
Jordá, Ó., Schularick, M., & Taylor, A. M. (2019). *The total risk premium puzzle (tech. rep.)*. National Bureau of Economic Research.
Kahneman, D., & Tversky, A. (1979). Prospect theory: An analysis of decisions under risk. *Econometrica, 47* (2), 263–291.
Kendall, M. G., & Hill, B. A. (1953). The analysis of economic time- series-part i: Prices. *Journal of the Royal Statistical Society, 116* (1), 11–34.
Mankiw, G. N., & Zeldes, S. P. (1991). The consumption of stockholders and nonstockholders. *Journal of financial Economics, 29* (1), 97–112.
Mehra, R. (2003). The equity premium: Why is it a puzzle? *National Bureau of Economic Research - Working Paper*, (9512).
Mehra, R., & Prescott, E. C. (1985). The equity premium: A puzzle. *Journal of monetary Economics, 15* (2), 145–161.
Merton, R. C. (1969). Lifetime portfolio selection under uncertainty: The continuous-time case. *Review of Economics and Statistics, 51* (2), 247–257.
Merton, R. C. (1971). Optimum consumption and portfolio rules in a continuous-time model. *Journal of Economic Theory, 3* (4), 373–413.
Niederhoffer, V., & Osborne, M. F. M. (1966). Market making and reversal on the stock exchange. *Journal of the American Statistical Association, 61* (316), 897–916.

Peleg, D. (2014). *Fundamental models in financial theory*. MIT University Press.
Perold, A. F. (2004). The capital asset pricing model. *Journal of economic perspectives, 18* (3), 3–24.
Roberts, H. V. (1959). Stock-market "patterns" and financial analysis: Methodological suggestions. *The Journal of Finance, 14* (1), 1–10.
Saltari, E., & Travaglini, G. (2009). Behavioral portfolio choice and disappointment aversion: An analytical solution with "small" risks. In *Nonlinear dynamics in economics, finance and social sciences: Essays in honour of john barkley rosser jr.* (pp. 295–311). Springer.
Samuelson, P. A. (1969). Lifetime portfolio selection by dynamic stochastic programming. *Review of Economics and Statistics, 51* (2), 239–246.
Scholes, M. (1969). *A test of the competitive hypothesis: The market for new issues and secondary offerings* [Doctoral dissertation, Graduate School of Business, University of Chicago].
Sharpe, W. (1966). Mutual fund performance. *The Journal of business, 39* (1), 119–138.
Sharpe, W., Alexander, G. J., & Bailey, J. W. (1999). *Investments*. Pearson.
Shleifer, A. (2000). *Inefficient markets: An introduction to behavioural finance*. Oxford University Press.
Weil, P. (1989). The equity premium puzzle and the risk-free rate puzzle. *Journal of Monetary Economics, 24* (3), 401–421.
Working, H. (1934). A random-difference series for use in the analysis of time series. *Journal of the American Statistical Association, 29* (185), 11–24.

Consumption and Economic Growth

4

- In this chapter, we study the neoclassical *Ramsey* growth model.
- Consumption and investment decisions are made *simultaneously*.
- The model assumes that the utility function of a *benevolent planner* represents collective preferences.
- We then show that a *decentralized economy* achieves the same growth path.
- The model is in continuous time and solved by the *Hamiltonian* function, with and without technological progress.
- We present the extension of the model to include *endogenous* technological progress and *sustainable* growth.
- Finally, we analyze the role of nonrenewable resources, the impact of pollution on welfare, and the case of sustainable consumption.

4.1 Introduction

We focus here on consumption and investment. This relationship can be studied in either a *partial equilibrium* growth model, as in Solow (1957) or in a *general equilibrium* model, where saving and investment are consequences of the intertemporal consumption decisions of households (Ramsey 1928). This framework is known as the *Ramsey-neoclassical* growth model.

In the general equilibrium model, the consumer provides labor, consumes, and saves, maximizing his or her intertemporal utility. However, firms operate in a competitive market and purchase capital, labor, and technology to maximize present and future profits. This model, variously developed by Ramsey (1928), Fisher (1930), Cass (1965), and Koopmans (1965), excludes from the analysis the possibility of market failures (such as imperfect competition, credit rationing, asymmetric infor-

G. Travaglini et al., *Notes on Consumption Theory*, Classroom Companion: Economics,
https://doi.org/10.1007/978-3-031-54986-1_4

mation, or heterogeneous agents). However, it provides a benchmark of sorts for analyzing the effect of such imperfections on the functioning of the economic system and its *positive* and *normative* properties. We examine this in detail in the next sections.

4.1.1 Consumers

The Ramsey model is expressed *per capita*. The representative household problem is not much different from the consumer problem studied so far. Now, however, there are several new elements to consider since we assume that population grows at a rate of n, that is, $L_t = L_0 e^{(nt)}$, where L_0 is the level of initial employment.

We define per capita consumption as $c_t = C_t/L_t$. It determines the utility of each period t. We assume that

$$u^{'}(c_t) > 0 \text{ and } u^{''}(c_t) < 0 \text{ for } c > 0, \tag{4.1}$$

with the properties

$$\lim_{c_t \longrightarrow 0} u^{'}(c_t) = \infty \quad \text{and} \quad \lim_{c_t \longrightarrow \infty} u^{'}(c_t) = 0. \tag{4.2}$$

That said, given the intertemporal preference rate θ, the household must solve the following problem:

$$\max \int_0^T u(c_t) L_t e^{(-\theta t)} dt = \int_0^T u(c_t) L_0 e^{(nt)} e^{(-\theta t)} dt \tag{4.3}$$

$$= L_0 \int_0^T u(c_t) e^{-[(\theta - n)t]} dt, \tag{4.4}$$

where, to guarantee the convergence of the integral, $\theta - n > 0$. This problem can be further simplified by writing $\rho \equiv \theta - n$, posing $L_0 = 1$ and assuming that the time horizon is infinite $T \longrightarrow \infty$.

4.1.2 Firms

Assume that the production function has the form $Y_t = F(K_t, L_t)$, with constant returns to scale (CRS) and positive and decreasing marginal productivity of inputs. The variable Y_t is output, K_t represents capital stock, and L_t is labor employed in production. This function can be rewritten in terms of product per capita by dividing both sides of the expression by L_t. We denote by lowercase letters the ratio $y_t = Y/L$, equal to the average product of labor, and $k_t = K_t/L_t$, equal to the capital-labor ratio (capital intensity). The production function can then be written in its intensive form as follows:

$$\frac{Y_t}{L_t} \equiv y_t = F\left(\frac{K_t}{L_t}, 1\right) = f(k_t) \text{ with } f^{'}(k_t) < 0 \text{ e } f^{''}(k_t) < 0 \text{ for } k_t > 0, \quad (4.5)$$

with the properties

$$\lim_{k \longrightarrow 0} f^{'}(k_t) = \infty \quad \text{and} \quad \lim_{k \longrightarrow \infty} f^{'}(k_t) = 0. \quad (4.6)$$

These are known in the literature as *Inada conditions* (Uzawa 1963).

The product Y_t is divided between gross investment I_t and consumption, C_t; $Y_t = C_t + I_t$. The coefficient $0 < \delta < 1$ is the rate of capital depreciation. Therefore, the net investment $\dot{K} = dK/dt$ is given by

$$\begin{aligned} \dot{K}_t &= I_t - \delta K_t \\ &= Y_t - C_t - \delta K_t, \end{aligned} \quad (4.7)$$

which, dividing by L, becomes

$$\frac{\dot{K}_t}{L_t} = y_t - c_t - \delta k_t = f(k_t) - c_t - \delta k_t, \quad (4.8)$$

with $c_t = C_t/L_t$. The left-hand side can be further simplified by noting that

$$\dot{K}_t = \frac{dK}{dt} = \frac{d}{dt}(k_t L_t) = k_t \frac{dL}{dt} + L_t \frac{dk}{dt} = k_t n L_t + L_t \dot{k}, \quad (4.9)$$

where $n = \frac{dL/dt}{L}$ is the rate of population growth (which, for simplicity, we assume coincides with that of employment growth). Substituting the latter expression into Eq. 4.8 for $\dot{K}$, we obtain the following dynamic equation:

$$\dot{k}_t = f(k_t) - c_t - (n + \delta) k_t. \quad (4.10)$$

This expression, often referred in the literature as the Solow equation, describes how investment per capita is a function of output, consumption, and the so-called "necessary investment", $(n + \delta) k_t$. This latter is the minimum investment necessary to maintain a constant per capita capital stock, k, as the labor force grows at rate n and capital stock depreciates at the rate δ. Finally, note that the difference $f(k_t) - c_t$ is saving as a function of k_t.

4.1.3 Ramsey Problem: the Centralized Economy

Ramsey (1928) raised the question of how much a nation can maximize its *collective* intertemporal welfare. The dynamic model employed in this study has become the prototype for analyses of optimal resource allocation. The model presented here essentially follows the Ramsey analysis.

Suppose that a *central planner* (benevolent dictator) wants to maximize the welfare of the representative household. The crucial choice is to determine how much to *consume* and how much to *save* to accumulate the capital used in the production of new goods to be consumed (or invested again) in the future.

The problem of economic growth can thus be written as

$$\begin{cases} max & \int_0^\infty u(c_t)e^{-\rho t}dt \\ \text{s.t.} & \dot{k}_t = f(k_t) - c_t - (n+\delta)\, k_t \\ & k(0) = k_0 \text{ e } k_t, c_t \geqslant 0. \end{cases} \tag{4.11}$$

In this problem, c_t is the *control* variable, while k_t is the *state* variable. Therefore, the optimal solution requires choosing the path of c_t that maximizes intertemporal utility, given the constraint on capital accumulation $\dot{k}_t$. In other words, consumption depends on production y_t and production on capital accumulation, which eventually depends on the *intertemporal* choice between consumption and saving, that is, new investment. The interaction of these choices determines the growth path of the economy in the long run.

We can solve Eq. (4.11) using the Hamiltonian

$$H = u(c_t)e^{-\rho t} + \lambda_t \left[f(k_t) - c_t - (n+\delta)\, k_t\right], \tag{4.12}$$

where λ_t is the *costate* variable associated with the state variable k_t; equivalently, λ_t is the shadow value of an additional unit of capital at time t.

The function H is *nonlinear* in c_t. Indeed, the first additive component of H is an *increasing function*, but with decreasing returns, of c_t, as illustrated in Fig. 4.1; the second component is a *decreasing linear function* of c_t. From the graph, we can see that their algebraic sum generates a *concave* function with a maximum between c_0 and c_1. Therefore, to calculate the maximum of H, the following condition must be satisfied:

$$\frac{\partial H}{\partial c_t} = 0 : u^{'}(c_t)e^{-\rho t} - \lambda_t = 0 \tag{4.13}$$

or

$$u^{'}(c_t)e^{-\rho t} = \lambda_t, \tag{4.14}$$

which states that the discounted per capita marginal utility, in each period, must equal the shadow price of capital. Moreover, since $\partial^2 H/\partial c_t^2 < 0$, the function H is

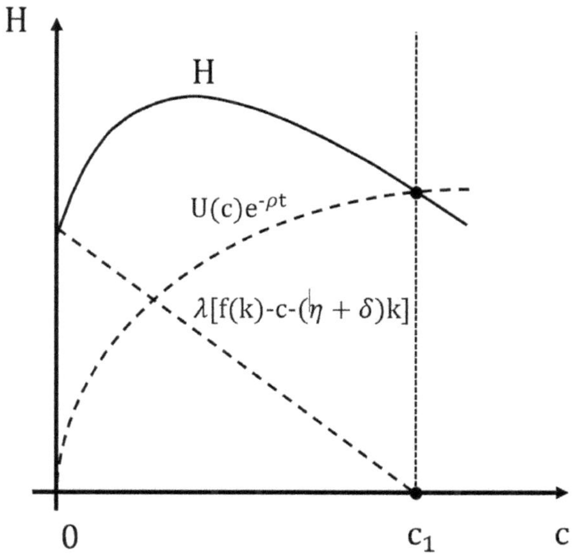

Fig. 4.1 The Hamiltonian function

maximized. However, as we know, the maximum principle requires the satisfaction of two additional dynamic conditions,

$$\dot{\lambda} = -\frac{\partial H}{\partial k} : \dot{\lambda} = -\lambda_t \left[f^{'}(k_t) - (n + \delta) \right], \tag{4.15}$$

$$\dot{k} = \frac{\partial H}{\partial \lambda} : \dot{k} = f(k_t) - c_t - (n + \delta) k_t, \tag{4.16}$$

and the condition of transversality, which takes the form

$$\lim_{t \longrightarrow \infty} k_t \lambda_t = 0. \tag{4.17}$$

The set of four equations, Eqs. 4.14–4.17 allows us to calculate the optimum values of the variables c_t, λ_t, and k_t. However, it is often preferred that the Hamiltonian be expressed in its present value. In this case, we write

$$H_c = u(c_t) + m_t \left[f(k_t) - c_t - (n + \delta) k_t \right], \tag{4.18}$$

where $H_c = He^{\rho t}$ and $m_t = \lambda_t e^{\rho t}$. In the latter reformulation, the optimal conditions are

$$u^{'}(c_t) = m_t, \tag{4.19}$$

$$\dot{m} = -\frac{\partial H_c}{\partial k} + \rho m_t : \dot{m} = -m_t \left[f^{'}(k_t) - (n + \delta + \rho) \right], \tag{4.20}$$

$$\dot{k} = \frac{\partial H_c}{\partial m} : \dot{k} = f(k_t) - c_t - (n + \delta) k_t. \tag{4.21}$$

In what follows, we use the *current-value Hamiltonian* to solve the optimal growth problem, Eq. 4.11.

4.1.4 The Modified Golden Rule

We begin by differentiating (4.19) with respect to time, thus obtaining an expression for $\dot{m}_t$ as follows:

$$u^{''}(c_t)\dot{c}_t = \dot{m}_t. \tag{4.22}$$

From this equation, together with $u^{'}(c_t) = m_t$ and Eq. 4.20 we get the following relationship:

$$\dot{c}_t = -\frac{u^{'}(c_t)}{u^{''}(c_t)}\left[f^{'}(k_t) - (n + \delta + \rho)\right], \tag{4.23}$$

which is a first-order differential equation in c_t. Equations 4.20 or 4.23 identify the Euler equation of the Ramsey problem, that is, the differential equation describing the necessary condition that must be satisfied by every optimal path. They represent, in continuous time, a relationship similar to that of Fisher (1930) that we consider in the first chapter, whereby the marginal rate of substitution must equal the marginal rate of transformation. This condition is known in the literature as the *Keynes-Ramsey rule*. It was derived by Ramsey in an original paper, the commentary to which is attributed to Keynes.

Therefore, to solve the problem, it is necessary to study the dynamics of the following system:

$$\begin{cases} \dot{k} = f(k_t) - c_t - (n + \delta) k_t \\ \dot{c}_t = -\frac{u^{'}(c_t)}{u^{''}(c_t)}\left[f^{'}(k_t) - (n + \delta + \rho)\right] \end{cases} \tag{4.24}$$

To this end, it is useful to represent the combination of forces influencing consumption (savings) and investment using the phase diagram.

4.1.5 The Phase Diagram

We first examine the properties of the system of differential equations (4.24). Technically, this is done by means of the phase diagram (see Appendix 1).

We must plot the two curves $\dot{k} = 0$ and $\dot{c}_t = 0$, that is

$$\begin{cases} c_t = f(k_t) - (n + \delta) k_t & \text{with } \dot{k} = 0 \\ f^{'}(k_t) = (n + \delta + \rho) & \text{with } \dot{c}_t = 0 \end{cases} \tag{4.25}$$

The first equation of the system has the same characteristics as the fundamental relation of the growth model in Solow (1957). It is a *concave* function in k_t, as a result of the difference between the per capita production function $f(k_t)$ (with decreasing returns in k_t) and the increasing line $(n + \delta) k_t$. Note that $\dot{k}_t = 0$ reaches its maximum when $f^{'}(k_t) = (n + \delta)$, that is, when the two functions have the same slope.

The second equation, $\dot{c}_t = 0$, implies that the *slope* of $f(k_t)$ is equal to $(n + \delta + \rho)$. Given the monotonicity of the production function, there is only one value of k^* for which this condition is satisfied. Therefore, as shown in Fig. 4.2, the graphical counterpart of the condition $\dot{c}_t = 0$ is an increasing straight line with a vertical intercept equal to k^*.

The intersection of the two curves $\dot{k} = 0$ and $\dot{c}_t = 0$ identifies the long-run equilibrium E, in which k and c no longer vary. In this equilibrium (*steady state*) c^* and k^* are the values of k and c that satisfy the so-called *modified golden rule*, considered below.

To understand the economic meaning of this condition, we note that the first equation of the system in (4.25) implies that $f^{'}(k_t) = (n + \delta)$ when $\dot{k} = 0$. It thus corresponds, in Fig. 4.2, to the point A on the curve $f(k_t)$ where the line tangent to the curve is parallel to the line $(n + \delta)\,k_t$. In contrast, k^* is defined by the condition $f^{'}(k_t) = (n + \delta + \rho)$, which implies, of course, a greater slope on $f(k_t)$ because $\rho > 0$. Therefore, the point B must lie to the left of A. Note that the value of c_t is a function of the value of k_t in the steady state. If ρ were zero, it would be sufficient to meet the condition $f^{'}(k_t) = (n + \delta)$. This condition is known in the literature as the *golden rule of capital* $\hat{k}$ to which corresponds the *golden rule of consumption* $\hat{c}$ that identifies the maximum per capita consumption given the capital stock in the steady state.

However, the *modified golden rule* states that this level of consumption is never attainable since the household's intertemporal preference, captured by the rate $\rho > 0$, requires that capital accumulation k_t be less than that set by the *golden rule* to satisfy intertemporal consumption preferences. In other words, although the household (and society) may consume more in the steady state corresponding to the golden rule of capital, the impatience of consumption implies that it is *suboptimal* to reduce current consumption to achieve the maximum consumption under the golden rule in the long run. Therefore, in the steady state (point B), both the capital stock k^* and consumption c^* that correspond to the modified gold rule will be smaller than the golden rule equivalents, that is, the pair of values $\left(\hat{k}, \hat{c}\right)$ of Fig. 4.2.

The *modified golden rule* is a relevant condition: it implies that capital productivity and real interest rate are ultimately influenced by the intertemporal preference rate, ρ. Therefore, consumption choices, the rates of population growth and capital depreciation determine the real interest rate $(n + \delta + \rho)$. Production is then the determinant of the long-run capital stock and consumption consistent with the real interest rate.

Optimal Control and Phase Diagram

We now focus on the diagram in Fig. 4.3. The curve $\dot{k} = 0$ begins at the origin of the axes, reaches a maximum at the *golden rule* level of capital $\hat{k}$ and cuts the horizontal axis at point A, where $f(k_t) = (n + \delta)\,k_t$. The curve $\dot{c}_t = 0$ is a vertical line, the value of which corresponds to the *modified golden rule*, k^*.

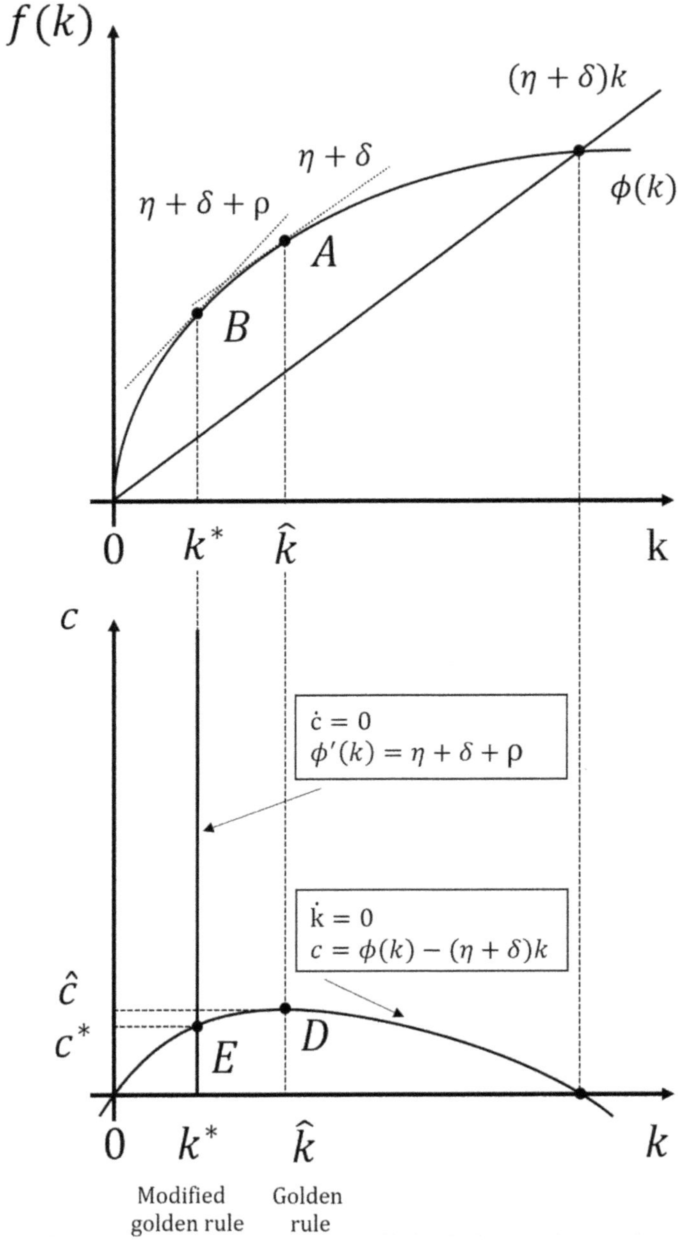

Fig. 4.2 Golden rule and modified golden rule

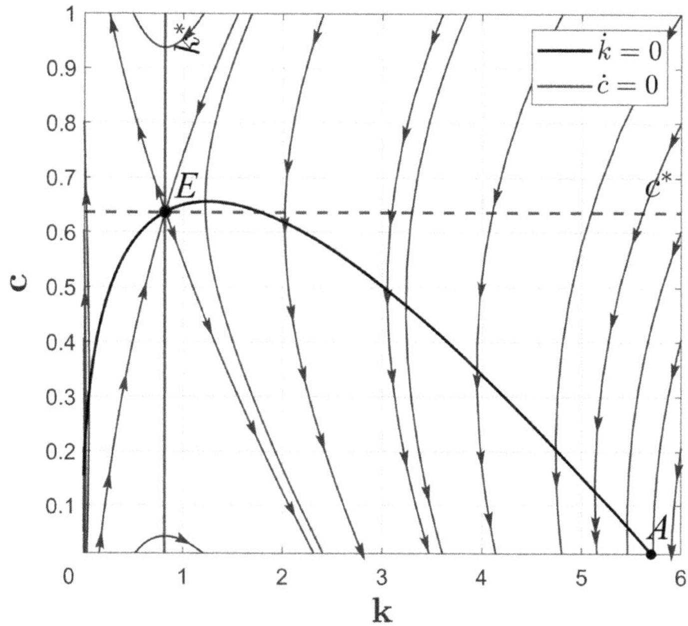

Fig. 4.3 The phase diagram

Above, $\dot{k} = 0$, k_t decreases continuously: consumption is, in fact, higher than that required to keep k_t constant. For the opposite reason, k_t increases below the $\dot{k} = 0$ curve.

The same applies to the $\dot{c}_t = 0$ curve. Consumption c_t increases to the left of the corresponding vertical line, where $f^{'}(k_t) > (n + \delta + \rho)$, and decreases to its right, where $f^{'}(k_t) < (n + \delta + \rho)$. The black arrows in Fig. 4.3 represent the direction of motion of the variables c_t and k_t. The system admits three distinct equilibria: the origin point, the point A, and the point E, defined by the intersection of the two curves. Note that the only way the economy can converge to the steady state E is along one of the trajectories leading to the long-run equilibrium. This means that, given the initial level of k_0, the economy must select the corresponding level of consumption, c_0, for which the pair (k_0, c_0) is *immediately* on the *stable branch* that converges to E.

If this condition is *not* satisfied, the dynamic forces in the model push the system toward divergent trajectories in which k_t increases continuously while c_t decreases, *or* c_t increases while k_t becomes smaller and smaller. Are these dynamics sustainable in the long run? Certainly not. In fact, the first scenario implies a continuous decrease in consumption, while in the second scenario, the increase in consumption eventually leads to the depletion of capital.

Since neither of these alternatives is sustainable, the *steady state* E, in which the level of capital and per capita consumption remain constant, is the only feasible long-run equilibrium of the Ramsey model. Once E is reached, equilibrium no longer changes.

This happens because the production function $y_t = f(k_t)$ is static: that is, there are no autonomous forces other than the capital-labor ratio that can alter y^* once the steady state k^* is reached.

Transversality Condition

An additional condition is necessary to get on the only stable path converging to equilibrium among all the possible trajectories crossing the phase diagram (k, c). This is the *transversality condition*, Eq. 4.17.

In the Ramsey model, the condition takes the form

$$\lim_{t \longrightarrow \infty} k_t \lambda_t = 0 \tag{4.26}$$

or simply

$$\lambda_t \rightarrow 0 \text{ for } t \rightarrow \infty. \tag{4.27}$$

This is because in the steady state $k^* > 0$ *but* the marginal value of an additional unit of capital must be zero; otherwise, it would still be convenient to exchange today's consumption for tomorrow's as determined by the Euler equation. In addition, the first-order condition

$$u^{'}(c_t)e^{-\rho t} = \lambda_t \tag{4.28}$$

implies that

$$\lim_{t \longrightarrow \infty} k_t u^{'}(c_t)e^{-\rho t} = 0 \tag{4.29}$$

since the discount factor is $\rho > 0$ and the horizon is infinite. In other words, in E the limit of $u^{'}(c^*)$ is finite for $t \rightarrow \infty$ and the expression in Eq. 4.29 satisfies the *transversality condition*.

Alternatively, it could be required that

$$H_t \rightarrow 0 \text{ when } t \rightarrow \infty. \tag{4.30}$$

Therefore, in our problem, the optimal solution for H takes the form

$$H^* = u(c^*)e^{-\rho t} + \lambda^* \left[f(k^*) - c^* - (n + \delta) k^* \right]. \tag{4.31}$$

Once again, since $u^{'}(c^*)$ is finite for $t \rightarrow \infty$, the exponential term $u(c^*)e^{-\rho t}$ tends to zero when t goes to infinity. As for the second term, λ^* tends to zero in the steady state because for values other than zero (recall that λ_t is the shadow price of capital), it would be optimal to continue to vary the capital stock (and consumption), thus contradicting the steady-state hypothesis. Moreover, in the *steady state*, $\dot{k} = 0$, which implies that $f(k^*) - c^* - (n + \delta) k^* = 0$. In short, in equilibrium, the transversality condition, Eq. 4.30, is satisfied.

The System Around the Steady State

In Fig. 4.2, two trajectories converge toward E, while two others diverge. When this happens, E is said to be a *saddle point*. In order to reach the steady state, it is necessary to select convergent paths that do not shift indefinitely away from E. Linearization of the dynamic system in (4.24) provides accurate information about the dynamic behavior of the economy near the long-run equilibrium. Linearizing both equations around the *steady state* (k^*, c^*) we get

$$\frac{\partial \dot{k}}{\partial k} = f(k^*) - (n + \delta) = \rho > 0 \tag{4.32}$$

$$\frac{\partial \dot{k}}{\partial c} = -1 < 0 \tag{4.33}$$

$$\frac{\partial \dot{c}}{\partial k} = -\frac{u^{'}(c^*)}{u^{''}(c^*)} f^{''}(k^*) < 0 \tag{4.34}$$

$$\frac{\partial \dot{c}}{\partial c} = \frac{-\left[u^{''}(c^*)\right]^2 + u^{'''}(c^*)U^{'}(c^*)}{\left[u^{''}(c^*)\right]^2} \left[f^{'}(k^*) - (n + \delta + \rho)\right] = 0 \tag{4.35}$$

from which we obtain the Jacobian matrix

$$J_E = \begin{bmatrix} \rho & -1 \\ -\frac{u^{'}(c^*)}{u^{''}(c^*)} f^{''}(k^*) & 0. \end{bmatrix} \tag{4.36}$$

The information on the characteristic roots v_1 and v_2 of the determinant of J_E needed to confirm the existence of a saddle is given by

$$v_1, v_2 = |J_E| = -\frac{u^{'}(c^*)}{u^{''}(c^*)} f^{''}(k^*) < 0, \tag{4.37}$$

from which it can be seen that the two roots have opposite signs and that the steady state of E is thus a *saddle point*.

4.1.6 The Decentralized Economy

Let us now suppose that the economy is decentralized. Can the steady-state change?

To describe this economy, we denote by r_t and w_t the return on capital and (real) wage, respectively. For simplicity, we assume that the rate of capital depreciation is zero; that is, $\delta = 0$. Firms maximize profits by remunerating capital and labor at their marginal productivity.

If, as in the centralized problem, the production function has the form $Y_t = F(K_t, L_t) = L_t f\left(\frac{K_t}{L_t}\right) = L_t f(k_t)$, with $k_t = \frac{K_t}{L_t}$, the two first-order conditions, with respect to K_t and L_t, are

$$F^{'}(K_t) = L_t f^{'}(k_t) \frac{1}{L_t} = f^{'}(k_t) = r_t \tag{4.38}$$

$$F^{'}(L_t) = f(k_t) - L_t f^{'}(k_t) \frac{K_t}{L_t^2} = f(k_t) - k_t f^{'}(k_t) = w_r \tag{4.39}$$

Now, the economy consists of a large number of identical households that are indifferent as to whether to hold their wealth in financial or real assets because interest rates on loans and capital are identical. The intertemporal problem becomes

$$\begin{cases} \max & \int_0^\infty u(c_t)e^{-\rho t} dt \\ \text{s.t.} & c_t + \frac{da_t}{dt} + na_t = w_t + r_t a_t \\ \text{where} & a_t \equiv k_t - b_t. \end{cases} \tag{4.40}$$

The constraint in Eq. 4.40 is the household's intertemporal budget, where a_t is the net wealth given by the capital, k_t, minus the debt b_t, w_t is labor income (wage) and $r_t a_t$ the financial return.

The maximization problem in Eq. 4.40, solved with the Hamiltonian, provides the following conditions

$$\dot{c}_t = -\frac{u^{'}(c_t)}{u^{''}(c_t)} \left[r_t - (n + \rho)\right] \tag{4.41}$$

$$\lim_{t \longrightarrow \infty} a_t u(c_t) e^{-\rho t} = 0 \tag{4.42}$$

as well as that of the intertemporal budget constraint.

Note that for the *No-Ponzi game condition*, in equilibrium, the private *aggregate* debt b_t must be zero so that, in the long run, $a_t = k_t$. Now, substituting Eqs. 4.38 and 4.39 into the intertemporal budget constraint Eq. 4.40, we obtain the following system of differential equations:

$$\begin{cases} \dot{k} = f(k_t) - c_t - nk_t \\ \dot{c}_t = -\frac{u^{'}(c_t)}{u^{''}(c_t)} \left[f^{'}(k_t) - (n + \rho)\right], \end{cases} \tag{4.43}$$

which describes the dynamic behavior of the decentralized economy. The two equations are identical to those of the centralized system (except for the depreciation rate δ assumed to be zero). Therefore, we find here, once again, the properties of the Ramsey model.

4.2 Technological Progress

The Ramsey model is characterized by a steady state in which output and consumption per capita are constant in the long run. Once this equilibrium is reached, the economy continues to reproduce itself in the same way. However, stylized facts and human history show that these variables have changed considerably over the centuries; in the last 200 years alone, per capita output and corresponding consumption have expanded extensively, contradicting the theoretical result of the model just outlined. Is it possible to overcome this impasse in the theory? How to make per capita economic variables "increase" in the steady state?

A notable feature of the model is the *static* nature of its production function. However, the basis of economic growth is technological change, and *technological progress* gradually changes the production function. Once technological progress enters the picture, one can easily overcome the conclusion that individuals' consumption and production are constant in the *steady state*. In the next section, we look at these features together.

4.2.1 A Dynamic Production Function

We can imagine that technological progress enters the production function $Y_t = F(K_t, L_t)$ in three different ways, of which only one is compatible with the Ramsey model.

Technological progress can be *Hicks-neutral* if the *marginal rate of transformation* (MRT)—that is, the ratio of the marginal productivities of capital and labor—is left unchanged. This definition can be understood using the following generalized form of the production function:

$$Y_t = A_t F(K_t, L_t), \tag{4.44}$$

where A_t is an exogenous variable that changes the scale of the output for the same inputs employed without changing the *MRT* as

$$MRT = \frac{A_t F_k}{A_t F_L} = \frac{F_k}{F_L}. \tag{4.45}$$

In contrast, technological progress is said to be *Harrod-neutral* if it leaves the capital-product ratio, K_t/Y_t, unchanged. In this case, the generalized production function takes the form

$$Y_t = F(K_t, A_t L_t), \tag{4.46}$$

where Y_t is linearly homogeneous in K_t and $A_t L_t$. Since A_t is multiplicative of labor L_t alone, this type of technology is called *labor-augmenting* as it improves labor efficiency. In this production process, labor and technology are perfect substitutes. Moreover, the fact that technological progress does not directly affect K_t explains why K_t/Y_t remains constant at the same marginal of capital in the presence of technological progress.

Finally, technological progress can be *Solow-neutral*, that is, it affects only capital. In this case, we write

$$Y_t = F(A_t K_t, L_t); \tag{4.47}$$

therefore, changes in A_t modify Y_t and the productivity of capital, but have no direct impact on labor L_t.

As we shall see below, only *Harrod-neutrality* is compatible with the Ramsey model in which output and consumption per capita grow steadily in equilibrium.

4.2.2 Ramsey Model with Technological Progress

Let us assume that the production function is $Y_t = F(K_t, A_t L_t)$. We define $A_t L_t$ as labor, measured in *efficiency units*, meaning that technological progress improves the productivity of labor, for example, through knowledge or education. For now, we assume that technology A_t grows over time at the constant (exogenously given) rate g. This is a strong assumption but useful for generalizing the Ramsey model. With this in mind, we redefine all per capita quantities in terms of efficiency units, that is

$$\tilde{y}_t = \frac{Y_t}{A_t L_t}, \ \tilde{k}_t = \frac{K_t}{A_t L_t} \text{ e } \tilde{c}_t = \frac{C_t}{A_t L_t}. \tag{4.48}$$

An example of a *labor-augmenting* production function is the Cobb-Douglas function

$$Y_t = F(K_t, A_t L_t) = K_t^{\alpha} (A_t L_t)^{1-\alpha} \tag{4.49}$$

whose intensive form is obtained by dividing both inputs by $A_t L_t$

$$\frac{Y_t}{A_t L_t} = f(\tilde{k}) = F\left(\frac{K_t}{A_t L_t}, 1\right) = \left(\frac{K_t}{A_t L_t}\right)^{\alpha} \tag{4.50}$$

$$f(\tilde{k}_t) = \tilde{k}_t^{\alpha} \tag{4.51}$$

with $f'(\tilde{k}) = \alpha \tilde{k}_t^{\alpha-1} > 0$ and $f''(\tilde{k}) = (\alpha - 1)\alpha \tilde{k}_t^{\alpha-2} < 0$. As in the Ramsey model, we assume that the initial level of capital and labor are given, and the initial level of technological progress is also known. Labor and technology grow at the relevant exogenous rates:

$$\dot{L}_t = nL_t; \quad \text{that is,} \quad L_t = L_0 e^{nt}. \tag{4.52}$$

and

$$A_t = gA_t; \quad \text{that is,} \quad A_t = A_0 e^{gt}. \tag{4.53}$$

Capital, K_t varies according to the rule

$$\dot{K}_t = \frac{dK}{dt} = \frac{d}{dt}\left(\tilde{k}_t A_t L_t\right) = \tilde{k}_t A_t \frac{dL}{dt} + L_t A_t \frac{d\tilde{k}}{dt} + \tilde{k}_t L_t \frac{dA}{dt}$$
$$= \tilde{k}_t A_t n L_t + L_t A_t \dot{k} + \tilde{k}_t L_t g A_t. \tag{4.54}$$

The evolution of $\dot{k}$ in efficiency units is thus given by the equation

$$\frac{\dot{K}_t}{A_t L_t} = \tilde{y}_t - \tilde{c}_t - \delta \tilde{k}_t = f(\tilde{k}_t) - \tilde{c}_t - \delta \tilde{k}_t \tag{4.55}$$

$$\tilde{k}_t n + \dot{k} + \tilde{k}_t g = \tilde{y}_t - \tilde{c}_t - \delta \tilde{k}_t = f(\tilde{k}_t) - \tilde{c}_t - \delta \tilde{k}_t \tag{4.56}$$

$$\dot{k} = f(\tilde{k}_t) - \tilde{c}_t - (g + n + \delta)\, \tilde{k}_t. \tag{4.57}$$

The optimal control problem then becomes

$$\begin{cases} \max & \int_0^\infty u(\tilde{c}_t) e^{-\rho t} dt \\ \text{s.t.} & \dot{k} = f(\tilde{k}_t) - \tilde{c}_t - (g + n + \delta)\, \tilde{k}_t \\ & \tilde{k}(0) = \tilde{k}_0 \text{ e } \tilde{k}_t, \tilde{c}_t \geqslant 0, \end{cases} \tag{4.58}$$

and the corresponding *current Hamiltonian* is

$$H_c = u(\tilde{c}_t) + m_t \left[f(\tilde{k}_t) - \tilde{c}_t - (g + n + \delta)\, \tilde{k}_t \right], \tag{4.59}$$

the optimality conditions are

$$u^{'}(\tilde{c}_t) = m_t, \tag{4.60}$$

$$\dot{m} = -m_t \left[f^{'}(\tilde{k}_t) - (g + n + \delta + \rho) \right], \tag{4.61}$$

and

$$\dot{k} = f(\tilde{k}_t) - \tilde{c}_t - (g + n + \delta)\, \tilde{k}_t \tag{4.62}$$

Solving the system and focusing on the equilibrium conditions for the new dynamic system $\dot{c} = 0$ and $\dot{k} = 0$, we derive the following pair of equations:

$$\begin{cases} \tilde{c}_t = f(\tilde{k}_t) - (g + n + \delta)\, \tilde{k}_t & \text{with } \dot{k} = 0 \\ f^{'}(\tilde{k}_t) = (g + n + \delta + \rho) & \text{with } \dot{c}_t = 0 \end{cases} \tag{4.63}$$

The only changes initially apparent are in the definitions of per capita variables and the algebra of the system, since g now appears in the two equations in (4.63). But, there is a drastic change to the modelâĿ™s properties, in particular those of the *steady state* $\tilde{k} = \tilde{k}^*$. By assumption, labor and technology grow at rates n and g, respectively. Since $\tilde{k}^*$ is constant in the steady state, the capital stock K_t must grow at the rate $(n + g)$ (i.e., $\frac{\dot{K}}{K} = n + g$). But, if, in equilibrium, capital and labor grow at $n + g$, Y_t will grow at the same rate, according to the CRS assumption. Therefore, K/L, C/L, and Y/L will grow at the rate of technological progress g; while capital, output, and consumption *in efficiency unit* will be constant. Since $\tilde{y}$ and $\tilde{c}$ are constant, the savings rate $\tilde{s} = \frac{(\tilde{y}-\tilde{c})}{\tilde{y}}$ will be constant as well."

Thus, the Ramsey model with technological progress implies that, regardless of the starting point, the economy converges to the *balanced growth path* where each variable grows at a constant rate, thereby overcoming the static nature of the neoclassical model. The main consequence of this extension is that, on the balanced growth path, the expansionary dynamics of output and per capita consumption are determined by the rate of technological progress, g.

4.2.3 Endogenous Technological Progress

An interesting question is whether the growth rate of technological progress, g, can change over time. Indeed, capital accumulation introduces the technology embedded in the inputs into the production process (Kaldor 1957, Solow 1960, Greenwood et al. 1997, Groth and Wendner 2014). Today's computers are more efficient than those of the past. Innovations change the technological constraints over time. Research and development acts similarly (Calcagnini et al. 2016a, 2016b). New technologies can have *constant* or *increasing* returns, rather than decreasing returns as in the Ramsey explanation.[1] In addition, education and schooling improve workers' knowledge and learning capacity, thus improving human capital. If these forces are at stake in the long run, technological progress can become cumulative, and the process of advancement becomes *endogenous*. How does the pattern of economic growth change with endogenous technological progress?

A simplified AK Model

The simplest model with *endogenous* technological progress is one in which output per capita is a linear function of the capital-labor ratio, that is, $Y_t = AK_t$. The key feature of this model, known in the literature as the *AK model*, is that, unlike the Ramsey model, capital productivity is *constant* over time.

We consider the intuition behind this formulation by assuming that $A = 1$, the depreciation rate $\delta = 0$, the population is constant ($n = 0$), and the utility function is the CRRA $u(c_t) = \frac{c_t^{1-\gamma}}{1-\gamma}$, where $c_t = C_t/L_t$. The problem then becomes

[1] Studies by Basu (1996) and Basu and Fernald (1997) suggest that returns to scale are constant or decreasing. Studies by Hall (1989), Caballero and Lyons (1992), Antweiler and Trefler (2002), and Harrison (2003), by contrast, suggest that there exist significant increasing returns, both internal and external.

$$\begin{cases} \max & \int_0^\infty (\frac{c_t^{1-\gamma}}{1-\gamma}) e^{-\rho t} dt \\ \text{s.t.} & \dot{k} = k_t - c_t, \end{cases} \tag{4.64}$$

because $y_t = k_t$. From the corresponding Hamiltonian, we obtain the optimal consumption rule:

$$\frac{\dot{c}}{c_t} = \frac{1}{\gamma} (1 - \rho), \tag{4.65}$$

and, after dividing the constraint in Eq. 4.64 by k_t, we can derive the equilibrium relationship between the rate of growth of capital per capita and the rate of growth of consumption per capita. More precisely,

$$\frac{\dot{k}}{k_t} = \frac{\dot{c}}{c_t} = 1 - \frac{c_t}{k_t} = \frac{1}{\gamma} (1 - \rho), \tag{4.66}$$

where $\frac{1}{\gamma}(1-\rho)$ represents the rate of economic growth. We know that along the balanced growth path, capital grows at the same rate as consumption. The relationship just derived shows this and states that the lower the growth, the higher the γ intertemporal elasticity of substitution and discount rate that determine the propensity to save.

We identify the relationship between the saving rate and economic growth by calculating s_t:

$$s_t = \frac{y_t - c_t}{y_t} = \frac{k_t - c_t}{k_t} = 1 - \frac{c_t}{k_t}, \tag{4.67}$$

which coincides with Eq. 4.66.

In short, if the economy does not grow, $\frac{\dot{k}}{k_t} = 0$ and savings are also zero. If $\frac{\dot{k}}{k_t} > 0$ also $s_t > 0$. In addition, since the ratio y_t/k_t is constant (in our example, equal to 1), the savings rate and the rate of capital accumulation have the same value. One consequence is that it is not possible to determine the direction of causality in the relationship between investment and saving because these are endogenous variables that depend on γ and ρ. In addition, the AK model creates a circle between saving and growth: an increase in saving stimulates investment, which increases output and consumption, simultaneously increasing saving. Thus, the model does not converge to a stationary equilibrium in the long run but tends to diverge over time. Therefore, compared to the Ramsey model, this model implies that the long-run growth rate is not constrained by the exogenous rate of technological progress g: an increased willingness to save results in an increasing level of capital, production, and consumption.

Figure 4.4 is the graphical counterpart of the analysis above. The term $s_t k_t$ represents total savings as a function of the capital stock. As we have seen, in this model,

Fig. 4.4 Savings and depreciation in the AK model

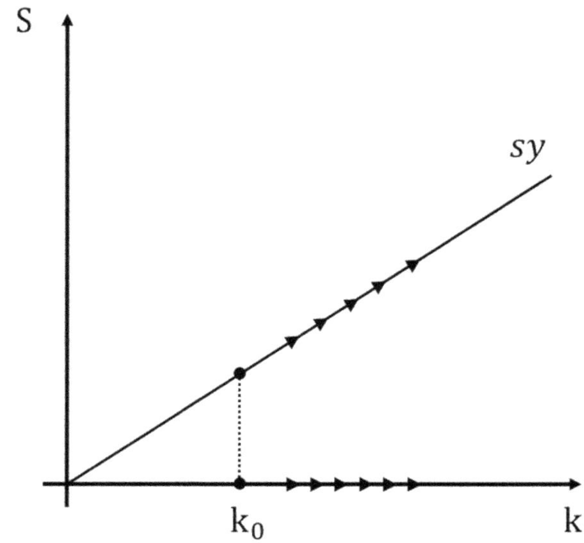

the savings rate is constant; therefore, given the linearity between y_t and k_t, the saving curve is also an increasing straight line. Consider an economy beginning with k_0. In this case, the capital stock grows over time and economic growth is unlimited.

Learning-by-Doing

Clearly, the result above is determined by the assumption of constant marginal productivity. However, less restrictive assumptions can provide more realistic scenarios, again characterized by *endogenous* technology. A pioneering model is the one presented by Kaldor (1957) and Kaldor and Mirrlees (1962), closely relating capital accumulation to the rate of productivity growth. The *technical progress function*, $\frac{dy}{y} = \alpha + \beta k$, captures this cause-and-effect relationship, making technological progress endogenous (McCombie and Spreafico 2015; Bellocchi et al. 2023).

In the same spirit, in a subsequent study Arrow (1962) proposes a microfounded model of *learning-by-doing*, assuming that the production function takes the form $Y_t = K_t^{\alpha} (A_t L_t)^{1-\alpha}$, but also that technological progress is influenced, in each period, by the experience gained in the working process; this depends, in turn, on the existing capital stock, K_t. This process makes technological progress *endogenous*. The function of technological progress is

$$A_t = B K_t^{\phi}, \tag{4.68}$$

with $B > 0$ and $\phi > 0$. Substituting Eq. 4.68 in the production function yields

$$Y_t = K_t^{\alpha+\phi(1-\alpha)} B^{1-\alpha} L_t^{1-\alpha}. \tag{4.69}$$

Now, the dynamics of the model depend on the fact that $\alpha + \phi (1 - \alpha) \gtreqless 1$, which implies the existence of increasing, constant, and decreasing returns to capital with direct impacts on production, accumulation, consumption, and, thus, on the dynamic properties of the model. A detailed presentation of these additional issues goes beyond the goals of consumption theory in this book. We therefore leave any further exploration to readers.

4.3 Sustainable Growth

Diminishing returns to scale or per capita resource scarcity are not the only factors that can limit economic growth. All human activity, consumption and production, inevitably generate some *pollution and waste*. Resources may be nonrenewable, like oil, or partially renewable (forests or fishes) and imply a high cost of restoration. In addition, their use and extraction can incur additional economic and environmental costs, in turn contributing to environmental degradation. There may be temperature changes as a result of increased soil, air, and water pollution, adversely affecting environmental quality, well-being of individuals, production of goods, and consumption choices. If economic data could include the environmental costs of economic growth, the use of the environment, in its various forms, would have to enter consumption and production choices with a negative price, influencing consumer (and social) utility, business decisions, and policy choices. Is it possible to quantify these *shadow prices*? And if so, how would they affect consumer and social welfare and intertemporal choices?

Hotelling (1931) posed this question, showing that when economic decisions are impacted by the use of natural resources and the quality of the environment, it is possible to identify an *environmental efficiency* condition for the use of natural resources and pollutant inputs; this condition closely relates the rate of society's intertemporal preference to the rate of change in the shadow value of natural resource *consumed* in production. This ecological-economic condition is known as the *Hotelling rule*.

4.3.1 The Hotelling's Rule

Suppose the objective function is to maximize the present value of the utility of households over the lifetime of a *nonrenewable resource* x_t, e.g., natural gas. The *social planner's problem* can be written as follows

$$\begin{cases} \max & \int_0^{\infty} u(c_t)e^{-\rho t}dt \\ \text{s.t.} & \dot{x} = -q_t \\ & \dot{k} = f(k_t, q_t) - c_t \end{cases} \tag{4.70}$$

The first constraint indicates that the stock of the *nonrenewable* resource x_t is reduced at the rate of extraction q_t, while the second constraint clarifies that the extracted resource is used in production (per capita) by an amount q_t equal to the flow $\dot{x}$ extracted in each period t.

Differently from the Ramsey model (without technological progress), we now have two state variables, k_t and x_t, and two control variables, c_t and q_t. Therefore, we will also need two costate variables, which we call m_t and p_t. The corresponding current *Hamiltonian* takes the form

$$H_c = u(c_t) + m_t [f(k_t, q_t) - c_t] + p_t (-q_t) \tag{4.71}$$

The necessary conditions for the maximum are

$$\begin{cases} c_t : u_c = m_t \\ q_t : m_t f_q = p_t \\ k_t : \dot{m} = -m_t f_k + \rho m_t \\ x_t : \dot{p}_t = \rho p_t \end{cases} \tag{4.72}$$

where u_c, f_q and f_k denote the first derivatives of each variable. Note that m_t and p_t are the *shadow values* of capital and nonrenewable resources, respectively. The last two equations of the system ensure compliance with the efficient dynamics. Dividing the equation related to capital by m_t gives

$$\frac{\dot{m}}{m_t} + f_k = \rho \tag{4.73}$$

This relationship states that along the optimal path, the return on capital, that is, the sum of capital gain $\frac{\dot{m}}{m_t}$ plus the marginal productivity f_k, must be equal to the social discount rate ρ. In short, we find the *Keynes-Ramsey rule*. Then, by dividing the equation concerning the shadow value of the nonrenewable resource by p_t, we get

$$\frac{\dot{p}_t}{p_t} = \rho. \tag{4.74}$$

This states that the rate of change of the environmental-resource shadow price must equal the social discount rate. This last equation is known as *Hotelling's rule*. This is an intertemporal efficiency condition that must be satisfied for any efficient resource extraction process. We encounter this condition (albeit with different formulations) in many economic-environmental analyses. A useful interpretation of this rule can be appreciated by rewriting Eq. 4.74 as $\dot{p}_t = \rho p_t$. The solution is $p_t = p_0 e^{\rho t}$, that is

$$p_o = p_t e^{-\rho t}, \tag{4.75}$$

from which it follows that p_0 is the discounted value at the rate ρ of the shadow price at time t. Hotelling's rule thus implies that there is no methodological difference between the pricing of financial or real assets and of an environmental resource. The environmental resource can be considered an *asset*, and its optimal use follows the usual intertemporal valuation rules related to consumption choices.

A final remark should be made. The efficient dynamics of the environmental price depends on the initial level p_0 of nonrenewable resources. Therefore, Hotelling's rule is a *necessary* but *not sufficient* condition for the optimal allocation of natural resources over time. It will be the global solution of the system reflected in Eq. 4.70 to provide the optimal paths of k_t, m_t, c_t, and p_t that maximize the welfare of the economy. In short, the rule offers a more complex solution but is otherwise no different from the problem originally addressed by Ramsey.

4.3.2 Consumption, Production, and Pollution

Pollution is caused by various emission sources. Consumption, production, and investment all generate pollution; it is considered a *by-product* of economic activity. Pollution is often caused by the use of fossil fuels, such as oil, coal, or natural gas. Therefore, it is reasonable to extend the previous model to the case where economic activities generate polluting emissions in the use of renewable or nonrenewable resources. The intertemporal consumption problem becomes even more complex than in the cases already explored. Consumers, firms, and society more generally have to control not only consumption and production but also the flow and stock of pollutants, which can, in turn, adversely affect *welfare*, consumption, and production. In short, *environmental sustainability* and *ecological transition* are crucial to economic decisions (Van Der Ploeg and Withagen 1991; Smith 2007; Calcagnini et al. 2016; Saltari and Travaglini 2016, 2017).

The Ecological-Economic System

We denote by $u(c_t, v)$ the per capita utility function, where v represents an *index* of *environmental pressure*; v has a *negative* effect on utility so $u_v < 0$ since it is assimilated to an environmental cost. The rate of environmental pressure v depends, in turn, on the rate of extraction of the nonrenewable resource $q_t \geqslant 0$ and the accumulated stock of pollution z_t (such as the total volume of greenhouse gases released into the atmosphere), which has a widespread impact on environmental quality and social well-being. We can then write $v = v(q_t, z_t)$, where $v_{q_t} > 0$ and $v_{z_t} > 0$. Under these assumptions, the consumption function becomes

$$u = u\left[c_t, v\left(q_t, z_t\right)\right]. \tag{4.76}$$

Of course, many forms of pollution are directly attributable to production, and pollution often has an adverse effect on production activity. The (per capita) production function can hence be generalized to include pollution-related environmental factors using the expression

$$y = f\left(k_t, q_t, v\left(q_t, z_t\right)\right) \tag{4.77}$$

where $f_k > 0$, $f_{q_t} < 0$ and $f_v < 0$. Furthermore, we assume that the extraction of the nonrenewable resource x_t has an increasing and convex cost,

$$g_t = g\left(q_t\right), \tag{4.78}$$

with $g_q > 0$ and $g_{qq} > 0$,; that is, resource extraction has a cost that increases progressively as extraction rates increase.

We complete our ecological-economic model by characterizing the dynamic of the relationship between pollutant emissions and investment. We begin with the pollution rate, which we write as

$$\dot{z}_t = n(q_t) - \omega z_t, \tag{4.79}$$

where $n(q_t)$ is the *gross* rate of pollution that depends positively on the rate of extraction q_t, thus $n_{q_t} > 0$, while $\omega > 0$ is the (exogenous) rate of natural pollution decay determined by the environmental system's natural ecological capacity to reabsorb pollutants over time.

Finally, regarding investment $\dot{k}$, we assume that in each period, expenditure of d_t can be incurred for finance environmental interventions that *abate* polluting emissions (*abatment activities*). As such, d_t is a component of the aggregate demand for which the following accounting identity holds true:

$$y_t = \dot{k} + c_t + g_t + d_t; \tag{4.80}$$

that is

$$\dot{k} = y_t - c_t - g_t - d_t. \tag{4.81}$$

Note that in the present model, the environmental expenditure d_t accelerates the pollution decay related to Eq. 4.79. However, the impact of this defensive expenditure on the pollution flux is not direct. Rather, it is determined by the efficiency of environmental technology, denoted by $h_t = h(d_t)$, where $h_d > 0$, such that the net flow of pollutant emissions, $\dot{z}_t$, is given by the equation

$$\dot{z}_t = n(q_t) - \omega z_t - h(d_t). \tag{4.82}$$

In summary, the pollution stock increases with emissions $n(q_t)$ but decreases as a result of natural decay ωz_t and environmental-defense activities, which have an impact equal to $h(d_t)$.

4.3.3 Sustainable Consumption

The dynamic optimization problem of sustainable consumption can be written as

$$\begin{cases} [c]lc_t : u_c = m_t \\ q_t : p_t = u_v v_q + m_t f_q + m_t f_v v_q - m_t g_q + \lambda_t n_q \\ d_t : m_t = -\lambda_t h_{d_t} \\ k_t : \dot{m} = -m_t f_k + \rho m_t \\ x_t : \dot{p}_t = \rho p_t \\ z_t : \dot{\lambda}_t = -u_v v_z - m_t f_v v_z + (\rho + \omega)\lambda_t \end{cases} \tag{4.83}$$

There are now three state variables k_t, x_t and z_t, three control variables c_t, q_t and d_t, and three costate variables, m_t, p_t and λ_t. The current value Hamiltonian can then be written as

$$\begin{aligned} H_c = {} & u\left[c_t, v\left(q_t, z_t\right)\right] + m_t\left[f\left(k_t, q_t, v\left(q_t, z_t\right)\right) - c_t - g\left(q_t\right) - d_t\right] \\ & + p_t\left(-q_t\right) + \lambda_t\left(n(q_t) - \omega z_t - h(d_t)\right), \end{aligned} \tag{4.84}$$

the optimum conditions of which are

$$\begin{cases} c_t : u_c = m_t \\ q_t : p_t = u_v v_q + m_t f_q + m_t f_v v_q - m_t g_q + \lambda_t n_q \\ d_t : m_t = -\lambda_t h_{d_t} \\ k_t : \dot{m} = -m_t f_k + \rho m_t \\ x_t : \dot{p}_t = \rho p_t \\ z_t : \dot{\lambda}_t = -u_v v_z - m_t f_v v_z + (\rho + \omega)\lambda_t \end{cases} \tag{4.85}$$

The three equations for c_t, k_t, and x_t are related to the optimization problems discussed above, and thus have the same meaning and interpretation. The analysis of the remaining three equations q_t, d_t, and z_t is somewhat more complicated.

The last equation of the system in (4.85) describes the efficient dynamics of the shadow price of pollution, λ_t, along the optimal path. We rewrite this relationship as

$$\lambda_t = \frac{u_v v_z + m_t f_v v_z}{(\rho + \omega)} + \frac{1}{(\rho + \omega)}\dot{\lambda}_t. \tag{4.86}$$

Equation 4.86 states that an economy concerned about its economic-ecological balance will, in each period t, devote specific resources to pollution control, up to the point where the marginal value of λ_t of the *utility* generated by pollution (i.e., the by-product of consumption and production) is equal to the marginal value—discounted at the intertemporal preference rate ρ plus the natural decay rate ω—of the pollution disutility, $u_v v_z + m_t f_v v_z$ due to the welfare and production consequences of emissions, net of the rate of change, $\dot{\lambda}_t$, of the shadow price of pollution, discounted at time t. In other words, the more a society wants to consume and produce, the more disutility it must be willing to endure to ensure this goal.

The second equation of the system in (4.85) provides the additional information needed to qualify Eq. 4.86. Notably, the equation

$$p_t = \left(m_t f_q - m_t g_q\right) + u_v v_q + m_t f_v v_q + \lambda_t n_q, \tag{4.87}$$

provides a measure of the environmental-resources shadow price. It shows that the price of an environmental resource is equal to the net marginal value $m_t f_q - m_t g_q$ of the resource (i.e., the value of the marginal product *net* of extraction costs) minus three *indirect* environmental costs:

1. $u_v v_q$, equal to the utility loss resulting from the impact on environmental pressure of a marginal unit of the nonrenewable resource extracted;

2. $m_t f_v v_q$, the loss of production from the use of a marginal unit of the nonrenewable resource due to the resulting environmental pressures that adversely affect production;
3. $\lambda_t n_q$, the value of the increase in marginal damage resulting indirectly from resource extraction as the nonrenewable stock declines;

These indirect damages tend to increase over time because a marginal increase in resource extraction results in an increase in emissions and, thus, in the level of environmental pollution. This ultimately affects welfare and production as a *negative externality*. To quantify this damage in economic terms, we need to transform it into the price of a unit of pollution.

Note, however, that in a centralized economy, such as that described here, these three pollution harms are *internalized*. This has obvious implications for *environmental policy*: for a decentralized economy in which the environmental cost caused by an agent's production or consumption activities generates negative externalities on society, an environmental agency should set an environmental *tax* equal to the value of the marginal pollution damage produced by extraction and production, that is, equivalent to $u_v v_q + m_t f_v v + \lambda_t n_q$.

This is the traditional welfare issue of negative *externalities* that occur whenever economic decisions affect the consumption or production decisions of other agents *outside* the market price channel. In the presence of externalities, the market is not Pareto-optimal. Correcting this distortion requires *property rights* to define contingent contracts between agents who have *internalized* the externality, either through a tax to reduce pollution or a subsidy to compensate for pollution.

Note, finally, that the third equation, $m_t = -\lambda_t h_{d_t}$, of the system in (4.85), implies that the marginal loss of value m_t for capital, determined by a shift of resources from investment production to the environmental activities $h(d_t)$, is optimal when the two values are equal at the margin. To understand this condition, let us recall the meaning of its terms. First, m_t is the *shadow price of capital*, or, in this context, the amount of utility foregone when a unit of production is diverted from consumption to expenditure on environmental defense. Second, λ_t is the *value* of a unit of *pollution*. Recall that this is a negative price since pollution is harmful. Third, h_{d_t} is the amount of *marginal pollution* abated by one unit of additional spending on environmental-defense activities. Therefore, the right-hand side of the equation, $-\lambda_t h_{d_t}$, is the (positive) value of the marginal utility obtained from pollution control when a unit of output is used for defensive spending. This marginal value must equal the value of utility given up by decreasing consumption (or investment) by one unit.

In other words, the optimal expenditure on pollution reduction will be the level at which the marginal costs and marginal benefits of pollution reduction are equal. In essence, it is a condition of environmental-economic efficiency where the *internalization* of costs allows the achievement of the social optimum. Therefore, the *Keynes-Ramsey rule* is extended to include the environmental costs and benefits of economic activity (Saltari and Travaglini 2017).

Appendix 1

4.3.3.1 Phase Diagram

The system in (4.25) in Sect. 4.1.5 comprises a pair of differential equations. These relations are often difficult to solve analytically, and a solution in closed form can be obtained only in (rare) cases. Fortunately, the *phase diagram* provides a simple and intuitive way to graphically represent the solution. Once the solution is identified, it is then possible to understand the structure of the dynamic system and study its properties.

In essence, we can divide the Cartesian space into regions where the value of variables increases or decreases. The *boundary conditions* will eventually provide constraints on the different trajectories, including those that are optimal for reaching the equilibrium point(s). In what follows, the logarithmic utility function is used to illustrate this technique.

Consider the problem of maximizing consumption c_t in a finite time interval $[0, T]$, as given below:

$$\begin{cases} \text{max.} & \int_0^T \ln(c_t)\exp(-\delta t)\,dt \\ \text{s.t.} & \dot{k}_t = k_t^\alpha - c_t \\ & k(0) = k_0 \\ & k(T) = k_T \end{cases} \tag{4.88}$$

where c_t denotes the consumption flow. In this formulation $k(t)$ is capital stock, $\dot{k}_t$ its variation (investment) and $y_t = k_t^\alpha$ is the production function whose output directly depends on k, and where $0 < \alpha < 1$. The logarithmic utility function is discounted at the subjective rate $\delta > 0$, and the boundary values $k(0)$ and $k(T)$ of the state variable are exogenously specified. The Hamiltonian of the problem is

$$H_t = \ln(c_t)\exp(-\delta t) + \lambda_t\left(k_t^\alpha - c_t\right). \tag{4.89}$$

By applying the maximum principle, we get

$$\frac{\partial H}{\partial c} = 0 : c_t^{-1}\exp(-\delta t) = \lambda_t \tag{4.90}$$

$$-\frac{\partial H}{\partial k} = \dot{\lambda} : -\lambda_t \alpha k_t^{\alpha-1} = \dot{\lambda} \tag{4.91}$$

$$\frac{\partial H}{\partial \lambda} = \dot{a} : \dot{k}_t = k_t^\alpha - c_t. \tag{4.92}$$

Let us use the first-order condition, Eq. 4.90 to eliminate c_t and obtain the following dynamic system

$$\begin{cases} \dot{\lambda} = -\lambda_t \alpha k_t^{\alpha-1} \\ \dot{k}_t = k_t^\alpha - e^{-\delta t}\lambda_t^{-1} \end{cases} \tag{4.93}$$

These differential equations involve not only λ and k but also the exponential time trend $\exp(-\delta t)$. This makes it impossible to draw a phase diagram in the space (k, λ) because the locus of points at which $\dot{k}_t = 0$ is not well defined since it depends on t.

Therefore, it is convenient to examine another pair of differential equations, one in k_t and one in c_t. To this end, we differentiate the condition in Eq. 4.90 with respect to time, obtaining

$$\begin{aligned} -\delta \exp(-\delta t)\, c_t^{-1} - \exp(-\delta t)\, c_t^{-2}\dot{c} &= \dot{\lambda} \\ &= -\lambda_t \alpha k_t^{\alpha-1} \\ &= -c_t^{-1} \exp(-\delta t)\, \alpha k_t^{\alpha-1}. \end{aligned} \tag{4.94}$$

Multiplying by $\exp(-\delta t)\, c_t^2$ gives

$$-\delta c_t - \dot{c} = -c_t \alpha k_t^{\alpha-1}, \tag{4.95}$$

from which we get

$$\dot{c} = c_t \left(\alpha k_t^{\alpha-1} - \delta\right). \tag{4.96}$$

Equation 4.96 is a differential equation in c_t, involving only c_t and k_t. Together, Eqs. 4.96 and 4.92, form the dynamic system

$$\begin{cases} \dot{c} = c_t \left(\alpha k_t^{\alpha-1} - \delta\right) \\ \dot{k} = k_t^{\alpha} - c_t, \end{cases} \tag{4.97}$$

which can now be analyzed using a phase diagram in the space (k, c).

The first step is to obtain the *loci* of the points where $\dot{k} = 0$ and $\dot{c} = 0$. The condition in Eq. 4.92 is satisfied along the curve $\dot{k}_t = k_t^{\alpha} - c_t = 0$, that is, $c_t = k_t^{\alpha}$. In other words, if output y_t is entirely consumed, savings, s, is zero, and investment, $\dot{k}$, is also zero. Since by assumption $\alpha < 1$, the function $c_t = k_t^{\alpha}$ is a concave and increasing curve in k. The curve starts at the origin point, where it has an infinitely large slope.

The second equation is obtained from the condition $\dot{c} = c_t \left(\alpha k_t^{\alpha-1} - \delta\right) = 0$, which means that $k^* = \left(\frac{\delta}{\alpha}\right)^{\frac{1}{\alpha-1}}$ is a constant value marked by k^*. The two critical *loci* are drawn in Fig. 4.5. They define four regions labeled counterclockwise, I-IV. In the positive space of the phase diagram, both k and c are positive, but changes to these can be positive or negative, depending on the initial conditions (starting point).

In Region I, c is less than k_t^{α}, hence the saving is positive and, consequently, $\dot{k}_t > 0$. Since $c > 0$ but $k < k^*$, the first equation of the system in (4.97) implies that even $\dot{c} > 0$. In Region II, we still have $c_t < k_t^{\alpha}$ and $\dot{k}_t > 0$. Now, however, $k > k^*$, so $\dot{c} < 0$. Region III is above the locus $c_t = k_t^{\alpha}$, so $\dot{k}_t < 0$ and $k > k^*$ imply $\dot{c} < 0$. In Region IV, $\dot{k}_t < 0$ and $\dot{c} > 0$ since for any $k < k^*$ the variable c increases while k decreases.

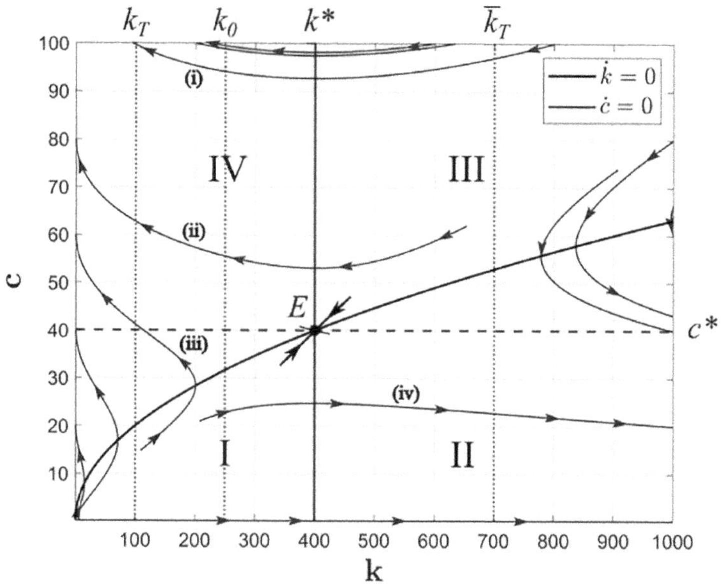

Fig. 4.5 Example of a phase diagram

The signs of $\dot{k}_t$ and $\dot{c}$ are represented in any region of Fig. 4.5 by the direction of the arrows. Since the k axis is horizontal and the c axis vertical, a perfectly horizontal arrow in the diagram would indicate an increase (right-hand direction) or decrease (left-hand direction) in k; a perfectly vertical arrow would refer to a positive (upward) or negative (downward) change in c. Each path moving from a point in one of the four regions thus depends on the intensity of these relative changes and indicates the direction of feasible paths in these regions.

In mathematical terms, the inclination of the trajectories in the (k, c) space can be obtained from the relation $dc/dk = (dc/dt) \,/\, (dk/dt) = \dot{c}/\dot{k}$. Therefore, when a trajectory passes through a point where $\dot{c} = 0$, it has a zero slope; by contrast, when it passes through a point where $\dot{k} = 0$, it has an infinite slope. Finally, the boundary values of the state variable k will determine the solution of the problem. We describe some possible paths that satisfy the maximum principle for the problem.

We denote by (k^*, c^*), the equilibrium values of the variables, since where the two loci cross $\dot{k} = \dot{c} = 0$, we have $k^* = \left(\frac{\delta}{\alpha}\right)^{\frac{1}{\alpha-1}}$ and $c^* = \left(\frac{\delta}{\alpha}\right)^{\frac{\alpha}{\alpha-1}}$. These define the equilibrium of the dynamic system. This long-run equilibrium is referred to as the *steady state*.

From our previous discussion, the feasible trajectories cover the entire (k, c) space. Hence, there is one trajectory in Region I that reaches the equilibrium E. We refer to this trajectory as a *stable* path since it converges to the steady state. Similar reasoning reveals the existence of a stable downward path in Region III, while in both Regions II and IV, there is an *unstable* path away from equilibrium, identifying E as a *saddle point*.

Suppose now that $k(0) = k_0$ and $k(T) = k_T$. The vertical dotted line in Fig. 4.5 identifies the boundary values of k. This means that the optimal trajectory must begin at a point along the line $k(0) = k_0$ and must end at some point along the line $k(T) = k_T$. Therefore, the optimal path will lie within Regions I and IV. In fact, there is no way to reach Region III when starting from k_0; furthermore, if the path enters Region II, it would never reach k_T. Note, however, that the optimal path can be entirely in Region IV, as with trajectories (i) and (ii), or start in Region I, as with (iii).

The choice of path depends on the time horizon $[0, T]$. Therefore, to solve the *optimal* problem, we must select the trajectory that will go from k_0 to k_T in exactly T units of time. The farther the path is from the $\dot{k} = 0$ curve, the larger $\dot{k}$ is, and the faster k will change. Thus, if T is small, the optimal path will be a high path, such as(i). The larger T is, the further down we go until we reach path (ii). For a very large T, we can select a path, such as (iii), that begins in Region I. Note that in the proximity of the steady state, E variables adjust very slowly. In short, if we expect it will take a long time to go from k_0 to k_T, as is the case along path (iii), it will be optimal to start with sufficiently low consumption to accumulate capital stock in the initial period. However, if time is short and $k_0 > k_T$, it is necessary to rapidly consume the initial capital stock k_0 to obtain k_T at time T.

Of course, alternative boundaries would provide different optimal paths. For example, suppose that the final capital stock must be higher than k_0, as it is for the value $\bar{k}_T$ in the graph. In this case, we could follow the path shown by (iv). Along this trajectory, k increases continuously, while c increases initially to decrease after a certain period in order to achieve $\hat{k}_T$ at time T. The time horizon $[0, T]$ would determine the optimal path since, as before, the closer the path is to $\dot{k} = 0$, the smaller the rate of change of k.

Lastly, it is also possible to drive the economy toward the *steady state*, E. However, to achieve the long-run equilibrium E, we must include in the present problem certain transversality conditions, such as an infinite horizon or some specific boundary conditions for the state variable. We discussed this specific aspect in our examination of the Ramsey model earlier in this chapter.

Further Reading

The Ramsey problem, with and without uncertainty, is well analyzed in Blanchard and Fischer (1989) (Chap. 2), Barro and Sala-i-Martin (1995) (Chap. 2), Ljungqvist and Sargent (2018) (Chap. 16) and Romer (2019) (Chap. 2). Barro and Sala-i-Martin (1995) (Chap. 3) also discuss some interesting extensions of the standard neoclassical model (government spending, installation costs for capital, open economy, and finite lifetime). A discussion of the mathematical properties of Ramsey model can be found in Gandolfo (2010) (Chap. 22). For technological change and endogenous growth, see instead Barro and Sala-i-Martin (1995) (Chap. 4), Acemoglu (2009) (Chaps. 11 and 12) and Aghion et al. (2014) (Chap. 1). On the relationship between human welfare and sustainability we suggest Heal (1998) while discounting future consumption with scarce resources is rigorously studied in Dasgupta (2001). Finally, a broad range of applications of dynamic optimization to issues of environmental economics (including the optimal extraction of resources) is presented in Moser et al. (2014) (Part IV) and (Perman 2003) (Part IV).

References

Acemoglu, D. (2009). Introduction to modern economic growth. Princeton University Press.

Aghion, P., Akcigit, U., & Howitt, P. (2014). Handbook of economic growth. North Holland.

Antweiler, W., & Trefler, D. (2002). Increasing returns and all that: A view from trade. American Economic Review, 92 (1), 93–119.

Arrow, K. J. (1962). The economic implications of learning by doing. The review of economic studies, 29 (3), 155–173.

Barro, R. J., & Sala-i-Martin, X. (1995). Economic growth. MIT University Press.

Basu, S. (1996). Procyclical productivity: Increasing returns or cyclical utilization? Quarterly Journal of Economics, 111 (3), 719–751.

Basu, S., & Fernald, J. G. (1997). Returns to scale in us production: Estimates and implications. Journal of political economy, 105 (2), 249–283.

Bellocchi, A., Travaglini, G., & Vitali, B. (2023). How capital intensity affects technical progress: An empirical analysis for 17 advanced economies. Metroeconomica.

Blanchard, O. J., & Fischer, S. (1989). Lectures on macroeconomics. MIT University Press.

Caballero, R. J., & Lyons, R. K. (1992). External effects in us procyclical productivity. Journal of Monetary Economics, 29 (2), 209–225.

Calcagnini, G., Favaretto, I., Giombini, G., Perugini, F., & Rombaldoni, R. (2016a). The role of universities in the location of innovative start-ups. The Journal of Technology Transfer, 41, 670–693.

Calcagnini, G., Giombini, G., Liberati, P., & Travaglini, G. (2016b). A matching model of university-industry collaborations. Small Business Economics, 46, 31–43.

Calcagnini, G., Giombini, G., & Travaglini, G. (2016). Modelling energy intensity, pollution per capita and productivity in italy: A structural var approach. Renewable and Sustainable Energy Reviews, 59, 1482–1492.

Cass, D. (1965). Optimum growth in an aggregative model of capital accumulation. Review of Economic Studies, 32 (3), 233–240.

Dasgupta, P. (2001). Human well-being and the natural environment. Oxford University Press.

Fisher, I. (1930). The theory of interest as determined by impatience to spend income and opportunity to invest it. The Macmillan Company.

Gandolfo, G. (2010). *Economic dynamics*. Springer.

Greenwood, J., Hercowitz, Z., & Krusell, P. (1997). Long-run implications of investment-specific technological change. American Economic Review, 87 (3), 342–362.

Groth, C., & Wendner, R. (2014). Embodied learning by investing and speed of convergence. Journal of Macroeconomics, 40, 245–269.

Hall, R. E. (1989). Invariance properties of solow's productivity residual. National Bureau of Economic Research Cambridge - Working Paper, (3034), 1–64.

Harrison, S. G. (2003). Returns to scale and externalities in the consumption and investment sectors. Review of Economic Dynam- ics, 6 (4), 963–976.

Heal, G. F. (1998). Valuing the future: Economic theory and sustainability. Columbia University Press.

Hotelling, H. (1931). The economics of exhaustible resources. Journal of Political Economy, 39 (2), 137–175.

Kaldor, N. (1957). A model of economic growth. The economic journal, 67 (268), 591–624.

Kaldor, N., & Mirrlees, J. A. (1962). A new model of economic growth. Review of Economic Studies, 29 (3), 174–192.

Koopmans, T. (1965). On the concept of optimal economic growth. In The econometric approach to development planning. North- Holland.

Ljungqvist, L., & Sargent, T. J. (2018). Recursive macroeconomic theory. MIT University Press.

McCombie, J. S., & Spreafico, M. R. (2015). Kaldor's 'technical progress function'and verdoorn's law revisited. Cambridge Journal of Economics, 40 (4), 1117–1136.

Moser, E., Semmler, W., Tragler, G., & Veliov, V. M. (2014). Dynamic optimization in environmental economics. Springer.
Perman, R. (2003). Natural resource and environmental economics. Pearson Education.
Ramsey, F. P. (1928). A mathematical theory of saving. The Economic Journal, 38 (152), 543–559.
Romer, D. (2019). Advanced macroeconomics, 5th edition. McGraw-Hill Esducation.
Saltari, E., & Travaglini, G. (2016). Pollution control under emission constraints: Switching between regimes. Energy Economics, 53, 212–219.
Saltari, E., & Travaglini, G. (2017). Optimal waste control with abatement capital. Journal of Evolutionary Economics, 27 (5), 1157–1180.
Smith, V. L. (2007). Rationality in economics: Constructivist and ecological forms. Cambridge University Press.
Solow, R. M. (1957). Technical change and the aggregate production function. The Review of Economics and Statistics, 39 (3), 312–320.
Solow, R. M. (1960). Investment and technical progress. In K. J. Arrow, S. Karlin, & P. Suppes (Eds.), Mathematical methods in the social sciences. Stanford University Press.
Uzawa, H. (1963). On a two-sector model of economic growth ii. The Review of Economic Studies, 30 (2), 105–118.
Van Der Ploeg, F., & Withagen, C. (1991). Pollution control and the ramsey problem. Environmental and Resource Economics, 1, 215–236.

Consumption and Uncertainty in Continuous Time

5

- The intertemporal consumption model is extended to the case of *uncertainty* in *continuous time*.
- Stochastic variables are described by *Wiener processes* and *Brownian motions*.
- We study the properties of these processes and show that although they are not differentiable in time, it is possible to use an approximation known as *Ito's Lemma*.
- This Lemma makes it possible to solve the stochastic HJB equation in continuous time.
- We derive more general implications for *stochastic optimal control* of consumption and saving.
- We conclude the chapter by applying these methods to five examples of stochastic consumption.

5.1 Introduction

In this chapter, we introduce stochastic control theory. We begin with an illustration of Wiener processes and Brownian motion, which are essential components for modeling shocks in continuous time. We introduce this concept with two different approaches: as a limit of random paths and as a Markov process. Then, we define stochastic integrals and discuss stochastic differential equations, examining the famous Ito's Lemma. In the course of the chapter, we warn the reader that some of the results and intuitions of previous chapters may no longer be true in stochastic calculus.

The second part of the chapter is devoted to stochastic optimization methods and their applications. We extend the HJB equation to the stochastic control problems and

G. Travaglini et al., *Notes on Consumption Theory*, Classroom Companion: Economics,
https://doi.org/10.1007/978-3-031-54986-1_5

study a set of sufficient conditions, including the transversality condition, to obtain optimal control. Finally, we apply this method to many examples, such as optimal consumption and portfolio rules, exhaustible resources, consumption of durables and credit rationing (Calcagnini and Saltari 2000).

5.2 Wiener Process or Brownian Motion

We know that a stochastic process describes the behavior of a variable that evolves over time according to a specific probability distribution. It is given by a set of random variables $[X_t, t \in T]$ in the probability space (Ω, f, p). In this notation, T is time, Ω is the space of states of the world, f the information set, and p is the probability of any of the events $[x_t \in X_t]$ in Ω. If T is countable, that is, $T = \{0, 1, 2, 3, \ldots\}$, the process is referred to as *discrete*. If, however, $T = [0, \infty)$, we say that T is not numerable and the process is *continuous*.

In Chap. 2, we studied the properties of the intertemporal consumption model with uncertainty under the assumption of *discrete* stochastic incomes and returns. We now generalize the model to the *continuous-time* case of a *Wiener's process or Brownian motion* $\{Z_t, t \in [0, \infty)\}$.

As will become clear, the problem is now much more complex since infinitesimal calculus with continuous stochastic variables differs in some important respects from stochastic differential calculus in discrete time. A rigorous treatment of such a problem requires an analytical exposition that goes far beyond the objectives of this book. Therefore, we present below a formal, but not rigorous, exposition of the problem at hand, which is, however, suitable for characterizing numerous economic contexts, including those of consumption. We begin by deriving the continuous stochastic motion, known as Brownian motion, starting from the properties of the *random walk* analyzed in the discrete case.

A *Wiener process or Brownian motion* must possess the following properties:

1. If $0 \leq t_0 < t_1 < \cdots < t_n < t$ is a temporal ordering, then for real values $x_0, x_1, \ldots$ x_n, the following relation must hold:

$$p\left[Z_t \leq x \mid Z_{t_0} = x_0, \ldots, Z_{t_n} = x_n\right] = p\left[Z_t \leq x \mid Z_t = x_n\right]. \tag{5.1}$$

   This means that Z_t is a Markovian process.
2. If $0 \leq t_0 < t_1 < \cdots < t_n < t$ are time points, then for any set of events in A

$$p\left[Z_t - Z_{t-i} \in A_i \text{ per } i \leq n\right] = \prod_{i \leq n} p\left[Z_t - Z_{t-i} \in A_i\right]. \tag{5.2}$$

   This means that the increments of the process $Z_t - Z_{t-i}$ are independent random variables.
3. For $0 \leq s \leq t$ the increment $Z_t - Z_{t-i}$ has distribution

$$p\left[Z_t - Z_{t-i} \in A_i\right] = \frac{1}{\sqrt{2\pi(t-s)}} \int \exp\left[\frac{-x^2}{2\,(t-s)}\right] dx; \tag{5.3}$$

that is, the increment $Z_t - Z_{t-i}$ is *normally* distributed with mean equal to 0 and variance equal to $\sigma^2(t-s)$. Note that, for the sake of simplicity, we assume in the above equation that $\sigma^2 = 1$. So, the Wiener process is a *non-stationary* stochastic process since the variance depends on the time interval $(t-s)$.

4. It is convenient to rewrite properties $1-3$ employing an alternative notation that is useful for our purposes. We posit $Z_t - Z_{t-i} = \Delta Z$ and $t - s = \Delta t$. In order for Z_t to follow a *Wiener process*, ΔZ must possess some additional properties:

   4.1 The relationship between ΔZ and Δt must be described by the equation

$$\Delta Z = \varepsilon_t \sqrt{\Delta t}, \tag{5.4}$$

   where ε_t is a *white noise* with mean equal to 0 and variance equal to 1.

   4.2 The variable ε_t is assumed to be serially uncorrelated with itself, that is, $E[\varepsilon_t, \varepsilon_s] = 0$ for $t \neq s$.

   4.3 The previous two conditions ensure the respect of the properties 1 and 2. In fact, it follows from Eq. 5.4, that ΔZ has a normal distribution with:

$$\begin{aligned} \text{mean: } & E[\Delta Z] = E\left[\varepsilon_t \sqrt{\Delta t}\right] = 0 \\ \text{variance: } & E[\Delta Z]^2 = E[\varepsilon_t]^2 \, E\left[\sqrt{\Delta t}\right]^2 = \Delta t; \end{aligned} \tag{5.5}$$

   that is, Z_t is a Markovian process.

Finally, let us consider the increment of Z_t over an extended interval, $Z_T - Z_0$. This interval can be interpreted as the sum of N small intervals of size Δt; that is, $N = T/\Delta t$. We can therefore write

$$Z_T - Z_0 = \sum_{i=1}^{N} \varepsilon_t \sqrt{\Delta t}, \tag{5.6}$$

from which it follows that

$$\begin{aligned} \text{mean: } & E[Z_T - Z_0] = 0 \\ \text{variance: } & E[Z_T - Z_0]^2 = N\Delta t = T. \end{aligned} \tag{5.7}$$

This result respects property 3 since the variance of the process increases with time T. In other words, the Wiener process or Brownian motion is *not stationary*.

As a final step, it should be remarked that in stochastic calculus, we refer to *infinitesimal* variations of the variable. Therefore, a Wiener process is the limit for $\Delta t \to 0$ of the process in Eq. 5.1. It can be rewritten as

$$dZ = \varepsilon_t \sqrt{dt}. \tag{5.8}$$

This latter equation is known as *Brownian motion*.

5.2.1 Generalized Wiener Process

So far, we have referred to a continuous stochastic process without additional deterministic elements. We now assume that the stochastic process develops around a deterministic trend (*drift*) and that the variance is 1. With these two assumptions, we obtain a new diffusion process referred to as the *Generalized Wiener* process or *absolute Brownian motion*, which we write as

$$dx = adt + \sigma dZ. \tag{5.9}$$

In this equation, a and σ are constants, and dZ is an increment of the Wiener process. The meaning of Eq. 5.9 can be understood by separately considering the two elements appearing on its right-hand side. Indeed, if $dZ = 0$ then Eq. 5.9 reduces to

$$dx = adt \Longrightarrow \frac{dx}{dt} = a \Longrightarrow x = x_0 + at, \tag{5.10}$$

and the variable x follows a deterministic continuous path with *drift* equal to a. However, when $dZ \neq 0$, the dynamic becomes stochastic, and dZ is a shock that shifts the variable x away from its trend, at. Substituting Eq. 5.8 into Eq. 5.9 we have

$$dx = adt + \sigma\varepsilon\sqrt{dt} \tag{5.11}$$

from which it follows that dx is normally distributed with

$$\begin{aligned} &\text{mean: } E\left[dx\right] = adt \\ &\text{variance: } E\left[dx\right]^2 = \sigma^2 dt. \end{aligned} \tag{5.12}$$

It follows that over a horizon of length T, the stochastic variable x is normally distributed with

$$\begin{aligned} &\text{average: } E\left[x\right] = aT \\ &\text{variance: } E\left[x\right]^2 = \sigma^2 T. \end{aligned} \tag{5.13}$$

The relationship between the Wiener process and the generalized Wiener process is shown in Fig. 5.1.

5.2.2 Generalized Brownian Motion or Ito Process

We can further generalize the previous stochastic process by assuming that the values a and σ are not parameters but functions of x and t. In this case Eq. 5.9 becomes

$$dx = a\left(x, t\right)dt + \sigma\left(x, t\right)dz. \tag{5.14}$$

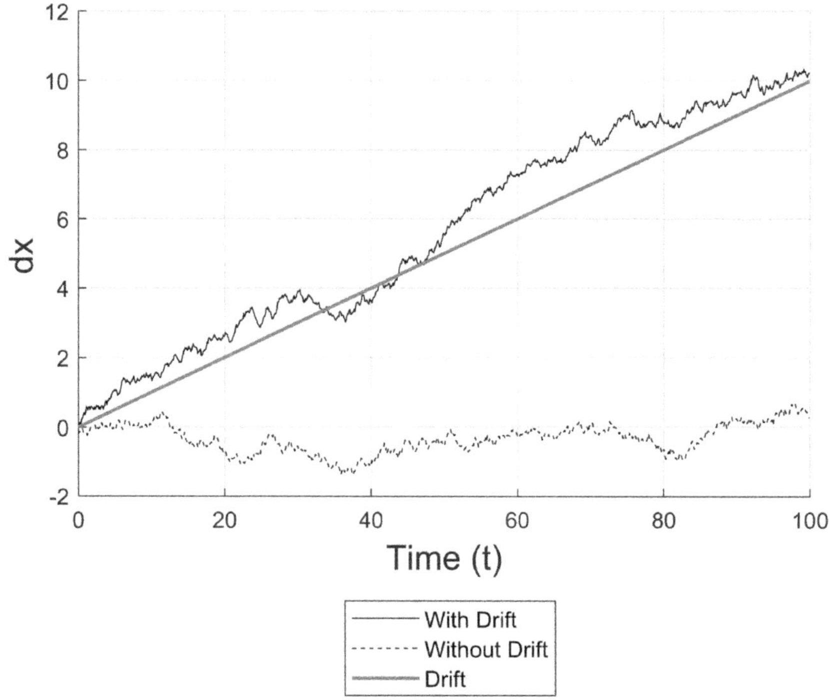

Fig. 5.1 Brownian geometric motion with and without drift

Such an equation is called *Ito process*. Again, dz is a Wiener process. Therefore, it follows that dx is normally distributed with

$$\begin{aligned} &\text{mean: } E\left[dx\right] = a(x,t)dt \\ &\text{variance: } E\left[dx\right]^2 = \sigma^2(x,t)dt. \end{aligned} \tag{5.15}$$

A special case of Eq. 5.14 is the *geometric* Brownian motion *with drift*. In this case we place $a\,(x,t) = ax$ and $\sigma\,(x,t) = \sigma x$, where, as before, a and σ are considered constant values. We get

$$dx = ax\;dt + \sigma x\;dz. \tag{5.16}$$

Finally, note that in the case of Eq. 5.16, the percentage variation of x, that is $\frac{dx}{x}$, is normally distributed. Since such fluctuations are variations of the natural logarithm of x, the absolute variations of x, that is dx are *log-normally distributed.*

Example 1.
Geometric Brownian motions are often used to describe stochastic movements in the stock prices or other traded *securities*, or to study changes in the value of portfolios. Here, we examine one such application.

Suppose that the current price of a security (*stock*) at time t is equal to S. If the expected rate of return a is constant, an investor will invest in the *stock* of value

S only if, in the interval dt, the expected return is equal to $aSdt$. Notably, if the variance of *stock price* were zero, Eq. 5.16 would be

$$dS = aS\,dt \Rightarrow \frac{dS}{S} = adt \Rightarrow S = S_0 e^{at}, \tag{5.17}$$

where S_0 is the price at time $t = 0$. However, the stock price is usually characterized by a certain degree of *volatility*. In our example, the total variance on a time interval dt is equal to $\sigma^2 S^2 dt$. Therefore, the instantaneous variance is $\sigma^2 S^2$ and the standard deviation is σS. Substituting into Eq. 5.16 we get the following stochastic expression of the *stock price* S:

$$\frac{dS}{S} = a\,dt + \sigma\,\varepsilon\sqrt{dt}. \tag{5.18}$$

Note that, again, $\frac{dS}{S}$ is normally distributed with

$$\begin{aligned} \text{mean: } & E\left[\frac{dS}{S}\right] = adt \\ \text{variance: } & E\left[\frac{dS}{S}\right]^2 = \sigma^2 dt. \end{aligned} \tag{5.19}$$

5.2.3 Non-differentiability of the Wiener Process

As mentioned above, the Wiener process Z_t is continuous in time and is *not differentiable*. A first way to show this is to observe that, from the third property for $0 \leq s \leq t$, it follows that

$$E\left(\frac{Z_t - Z_s}{t - s}\right)^2 = \frac{\sigma^2}{t - s} = \frac{1}{t - s}. \tag{5.20}$$

Assuming, as we did in Eq. 5.20, that $\sigma^2 = 1$, we get

$$\lim_{t \longrightarrow s} E\left(\frac{Z_t - Z_s}{t - s}\right)^2 = \infty. \tag{5.21}$$

However, if the Wiener process were differentiable, and Z'_s were its first derivative, then we should get

$$\lim_{t \longrightarrow s} E\left(\frac{Z_t - Z_s}{t - s} - Z'_s\right)^2 = 0 \Longrightarrow \lim_{t \longrightarrow s} E\left(\frac{Z_t - Z_s}{t - s}\right)^2 = E\left(Z'_s\right)^2 \tag{5.22}$$

which *contradicts* the previous result.

The *non-differentiability* of the process can also be shown using Eq. 5.9 directly. In this case, dividing by dt, we get

$$\frac{dx}{dt} = ax + \sigma x \frac{dZ}{dt} \tag{5.23}$$

which always contains the term

$$\frac{dZ}{dt} = \frac{\varepsilon\sqrt{dt}}{dt} = \varepsilon\,(dt)^{-1/2}\,. \tag{5.24}$$

Notably, this tends to *infinity* when $dt \to 0$.

5.3 Ito's Lemma

Although Wiener's process is not differentiable, it is possible to compute an approximation of it, known as *Ito's lemma* after the mathematician by whom it was first derived (Itô 1994, 1951).

Assume that the stochastic variable x follows the generalized Brownian motion

$$dx = a\,(x,t)\,dt + \sigma\,(x,t)\,dz, \tag{5.25}$$

and that x affects the stochastic process $G\,(x,t)$, differentiable in x and t. Taylor expanding the function $G\,(x,t)$, we obtain

$$\begin{aligned} dG = {} & \frac{\partial G}{\partial x}dx + \frac{\partial G}{\partial t}dt + \frac{1}{2}\frac{\partial^2 G}{\partial x^2}(dx)^2 \\ & + \frac{\partial G}{\partial x \partial t}dxdt + \frac{1}{2}\frac{\partial^2 G}{\partial t^2}(dt)^2 + \frac{1}{6}\frac{\partial^3 G}{\partial x^3}(dx)^3 + \cdots. \end{aligned} \tag{5.26}$$

In deterministic differential calculus, the terms of a higher order than the first tend to zero when x and t tend to zero. This implies that Eq. 5.26 should reduce to $dG = \frac{\partial G}{\partial x}dx + \frac{\partial G}{\partial t}dt$.

However, now dx follows a Brownian process. The element $(dx)^2$ is thus nonzero but equal to $\sigma^2\,(x,t)\,dt$, since

$$(dx)^2 = a^2\,(x,t)\,(dt)^2 + 2a\,(x,t)\,\sigma\,(x,t)\,(dt)^{3/2} + \sigma^2\,(x,t)\,dt. \tag{5.27}$$

Note how the terms $(dt)^2$ and $(dt)^{3/2}$ converge to zero faster than dt when this variation tends to zero. The first two terms can thus be ignored, and Eq. 5.27 reduces to

$$(dx)^2 = \sigma^2\,(x,t)\,dt. \tag{5.28}$$

Equation 5.26 can then be rewritten as

$$dG = \frac{\partial G}{\partial x}dx + \frac{\partial G}{\partial t}dt + \frac{1}{2}\frac{\partial^2 G}{\partial x^2}\sigma^2\,(x,t)\,dt, \tag{5.29}$$

which is an approximation of the stochastic calculus known as *Ito's lemma*. Finally, substituting dx (from Eq. 5.25 into Eq. 5.29) yields the expanded form of dG, that is,

$$dG = \left[a(x,t)\frac{\partial G}{\partial x} + \frac{\partial G}{\partial t} + \frac{1}{2}\sigma^2(x,t)\frac{\partial^2 G}{\partial x^2} \right] dt + \sigma^2(x,t)\frac{\partial G}{\partial x} dt. \tag{5.30}$$

This approximation applies to all the various Brownian motions studied so far. For example, for a stochastic process that depends on a *geometric* Brownian motion with constant drift and variance, *Ito's lemma* simplifies the situation. It allows us to ignore the derivatives of the drift and variance with respect to x and time t. But let us look at some examples.

Example 1.

1. Assume that x follows the geometric Brownian motion with constant a and σ

$$dx = ax\,dt + \sigma x\,dz. \tag{5.31}$$

We use Ito's lemma to show that the continuous stochastic process followed by a function $f(x) = \log x$ is equal to

$$\begin{aligned} df &= \frac{1}{x}dx - \frac{1}{2x^2}(dx)^2 \\ &= \alpha dt + \sigma dz - \frac{1}{2}\sigma^2 dt \\ &= \left(\alpha - \frac{1}{2}\sigma^2\right) dt + \sigma dz \end{aligned} \tag{5.32}$$

since $\partial f/\partial t = 0$, $\partial f/\partial x = 1/x$, and $\partial^2 f/\partial x^2 = -1/x^2$. Note that the drift of $f(x) = \log x$ is less than α. Why does this happen? The reason is that $\log x$ is a concave function of x, so with x stochastic, the expected value of $\log x$ changes *less than* the logarithm of the expected value of x. Finally, note that

$$E(df) = \left(\alpha - \frac{1}{2}\sigma^2\right) dt \tag{5.33}$$

$$E(df)^2 = \sigma^2 dt. \tag{5.34}$$

Example 2.

2. We now assume that $f(x) = e^x$. In this case, noting that $\partial f/\partial t = 0$, we get

$$\begin{aligned} df &= e^x(dx) + \frac{1}{2}e^x(dx)^2 \\ &= e^x(ax\,dt + \sigma x\,dz) + e^x\frac{1}{2}\sigma^2 x^2\,dt \end{aligned} \tag{5.35}$$

$$= e^x \left(ax + \frac{1}{2}\sigma^2 x^2 \right) dt + e^x \sigma x \, dz$$

$$\frac{df}{f} = \left(ax + \frac{1}{2}\sigma^2 x^2 \right) dt + \sigma x \, dz. \tag{5.36}$$

Example 3.

3. Assume that $C(x) = x^\theta$ with $\theta < 1$ is the consumption function. Using Ito's lemma, we obtain

$$\begin{aligned} dC &= \theta x^{\theta-1} (\alpha x \, dt + \sigma x \, dz) + \frac{1}{2}\theta (\theta - 1) x^{\theta-2} \sigma^2 x^2 \, dt \\ &= \left[\alpha\theta + \frac{1}{2}\theta (\theta - 1) \sigma^2 \right] C \, dt + \theta\sigma C \, dz. \end{aligned} \tag{5.37}$$

Then, it follows that

$$E\,(dC) = C \left[\alpha\theta + \frac{1}{2}\theta (\theta - 1) \sigma^2 \right] dt \tag{5.38}$$

$$E\,(dC)^2 = \theta^2 \sigma^2 C^2 dt. \tag{5.39}$$

Therefore, consumption behaves as a geometric Brownian motion. Moreover, recalling what has already been shown for the stock price S, we have that the expected utility of consumption in $t = 0$ is equal to

$$E\,[C(x)] = x_0^\theta e^{\left(\alpha\theta + \frac{1}{2}\theta(\theta-1)\sigma^2\right)t}. \tag{5.40}$$

Example 4.

4. Finally, we focus on the capital accumulation equation seen in the Ramsey problem, which we write as

$$\begin{aligned} \dot{K}_t &= I_t - \delta K_t \\ &= F(K_t, L_t) - C_t - \delta K_t \end{aligned}. \tag{5.41}$$

$$dK = [F(K_t, L_t) - C_t - \delta K_t] \, dt. \tag{5.42}$$

Now, we denote with lowercase letters the ratios $y_t = Y/L$ and $k_t = K_t/L_t$. The homogeneous production function becomes $F(\frac{K}{L}, 1) = f(k_t)$. Furthermore, we define per capita consumption as $c_t = C_t/L_t$. To introduce uncertainty in continuous time, we assume that the population dynamic is described by the following stochastic differential equation:

$$dL = nLdt + \sigma Ldz. \tag{5.43}$$

From Ito's lemma applied to $k_t = K_t/L_t$, we get

$$
\begin{aligned}
dk = d\left(\frac{K}{L}\right) &= \frac{dK}{L} - \frac{K}{L^2}dL - \frac{(dK)(dL)}{L^2} + \frac{K}{L^3}(dL)^2 \\
&= \left[\frac{F(K_t, L_t) - C_t - \delta K_t}{L}\right]dt + \frac{K}{L}\frac{dL}{L} + \left(\frac{K}{L}\right)\left(\frac{dL}{L}\right)^2, \qquad (5.44) \\
&= [f(k) - c - \delta k]\,dt - k(nLdt + \sigma Ldz) + \sigma^2 kdt \\
&= \left[f(k) - c - (n + \delta - \sigma^2)k\right]dt - \sigma^2 kdz
\end{aligned}
$$

since $(dK)(dL) = o(dt)$. Note that when $\sigma^2 = 0$, we obtain the standard capital accumulation equation of the Ramsey model

$$dk = [f(k) - c - (n + \delta)\,k]\,dt. \qquad (5.45)$$

5.4 Stochastic Dynamic Programming

In this section, we introduce the main properties of *stochastic dynamic programming* or stochastic control as applied to well-known economic problems. We begin with a simple model and then extend the discussion to more complex cases.

5.4.1 Absolute Brownian Motion

Consider the following absolute Brownian motion:

$$dx = adt + \sigma dZ, \qquad (5.46)$$

and a continuous function $\pi(x)$, the expected present value of which is $F(x)$:

$$F(x) = E\int_0^\infty \pi(x)e^{-rt}dt, \qquad (5.47)$$

where r is the interest rate. It is possible to consider an alternative characterization of the problem in Eq. 5.47. To this end, we divide the integral of Eq. 5.47 into the initial *infinitesimal* contribution of $\pi(x)$ between 0 and dt plus the integral between dt and ∞. Actually, we apply in a continuous stochastic context the procedure already used in the section on *dynamic programming* (2.2.1). The only relevant difference is that now we do not have a specific control variable (such as consumption, which we will reintroduce later), but only the *state* variable x that moves following a stochastic dynamic.

At time dt, the variable x will have a value of $(x + dx)$, which is unknown at time 0, although its probability distribution is known. The integral between dt and future periods is denoted by $F(x + dx, t + dt)$. We can write that the integral in Eq. 5.47 is equivalent to

$$\begin{aligned} F(x) &= \int_0^{dt} \pi(x)dt + E\int_{dt}^{\infty} \pi(x + dx, t + dt)e^{-rt}dt \\ &= \pi(x)dt + E\left[F\left(x + dx, t + dt\right)\right]e^{-rdt}. \end{aligned} \tag{5.48}$$

Note that the first addend of the latter equation is already an approximation since $\pi(x)$ is considered constant on the infinitesimal interval dt. This is possible because the approximation error in $\pi(x)dt$ is of the order of dt^2, and therefore negligible when dt tends to zero. To simplify this equation further, recall that for infinitesimal variations of dt, we have that $e^{-rdt} = 1 - rdt + o(dt)$, where $o(dt)$ denotes the terms that zero faster than dt. Further, even for future values of $F\left(x + dx, t + dt\right)$, the terms less than dt are of negligible order when $dt \to 0$. Therefore, Eq. 5.48 can be rewritten as

$$\begin{aligned} F(x) &= \pi(x)dt + (1 - rdt)\left(F(x) + \{E\left[F\left(x + dx\right)\right] - F(x)\}\right) \\ &= \pi(x)dt + F(x) - rF(x)dt + \{E\left[F\left(x + dx\right)\right] - F(x)\} \end{aligned}; \tag{5.49}$$

that is,

$$rF(x)dt = \pi(x)dt + \{E\left[F\left(x + dx\right)\right] - F(x)\}. \tag{5.50}$$

We get

$$rF(x)dt = \pi(x)dt + E\left(dF\right), \tag{5.51}$$

which is the *HJB equation*. Finally, applying Ito's lemma to $E\left(dF\right)$ we get

$$E\left(dF\right) = aF^{'}(x)dt + \frac{1}{2}\sigma^2 F^{''}(x)dt, \tag{5.52}$$

and the differential equation in x

$$\frac{1}{2}\sigma^2 F^{''}(x) + aF^{'}(x) - rF(x) + \pi(x) = 0. \tag{5.53}$$

The General Solution.

Equation 5.53 is a second-order linear differential equation in x. Therefore, its *general solution* is the sum of two separate components: the *particular* solution and the *complementary* solution of its homogeneous part (with the omission of $\pi(x)$).

We start from the *complementary* solution by rewriting only the homogeneous part of the equation

$$\frac{1}{2}\sigma^2 F^{''}(x) + aF^{'}(x) - rF(x) = 0. \tag{5.54}$$

To solve Eq. 5.41, it is necessary to guess the functional form of the solution. We assume (as a solution) the function $F(x) = Ae^{\lambda x}$ with $F^{'}(x) = A\lambda e^{\lambda x}$ and $F^{''}(x) = A\lambda^2 e^{\lambda x}$. Substituting we get

$$Ae^{\lambda x}\left(\frac{1}{2}\sigma^2\lambda^2 + a\lambda - r\right) = 0. \tag{5.55}$$

Since the exponential is always positive, the condition above is satisfied *if and only if*

$$-\frac{\sigma^2}{2}\lambda^2 - a\lambda + r = 0. \tag{5.56}$$

The latter condition often occurs in similar approximations of the *Ito's lemma*. For this reason, it is convenient to employ a general notation to study its properties. To this aim, we denote by ξ a generic variable and define the function

$$Q(\xi) = -\frac{\sigma^2}{2}\xi^2 - a\xi + r. \tag{5.57}$$

Although the value of the two roots can be explicitly calculated, it is useful to represent them graphically. Figure 5.2 describes the parabola Q as a function of ξ. It is called the *quadratic fundamental equation of Brownian motion*. The coefficient of ξ_2 is negative and, therefore, $Q(\xi)$ converges to $-\infty$ as ξ tends to $\pm\infty$. Since, by hypothesis $r > 0$, we have that $Q(0) = r > 0$, and the quadratic equation $Q(\xi) = 0$ has two roots, one positive, $\xi_1 > 0$ and one negative $\xi_2 < 0$. Therefore, $Q(\xi)$ is positive when $\xi_2 < \xi < \xi_1$ and negative outside the interval.

To conclude, and returning to our problem, the condition of convergence of the integral in Eq. 5.47 requires that λ lies between the values of the two roots $\lambda_2 < 0$ and $\lambda_1 > 0$, and can take the value $\lambda = 0$.

Therefore, the *complementary* solution has the form

$$F(x) = A_1e^{\lambda_1 x} + A_2e^{\lambda_2 x}, \tag{5.58}$$

with the constants A_1 and A_2 to be determined according to the *boundary conditions* and the values at time $t = 0$ of the state variable $x(0) = x_0$.

As we come to the particular solution, the easiest function to guess is $\pi(x) = e^{\lambda x}$, which implies $F(x) = Ke^{\lambda x}$. Substituting the latter expression into Eq. 5.53 we have

$$K\left(\frac{1}{2}\sigma^2\lambda^2 + a\lambda - r\right)e^{\lambda x} + e^{\lambda x} = 0; \tag{5.59}$$

that is,

$$K = \frac{1}{r - \frac{1}{2}\sigma^2\lambda^2 - a\lambda} = \frac{1}{\phi(\lambda)}, \tag{5.60}$$

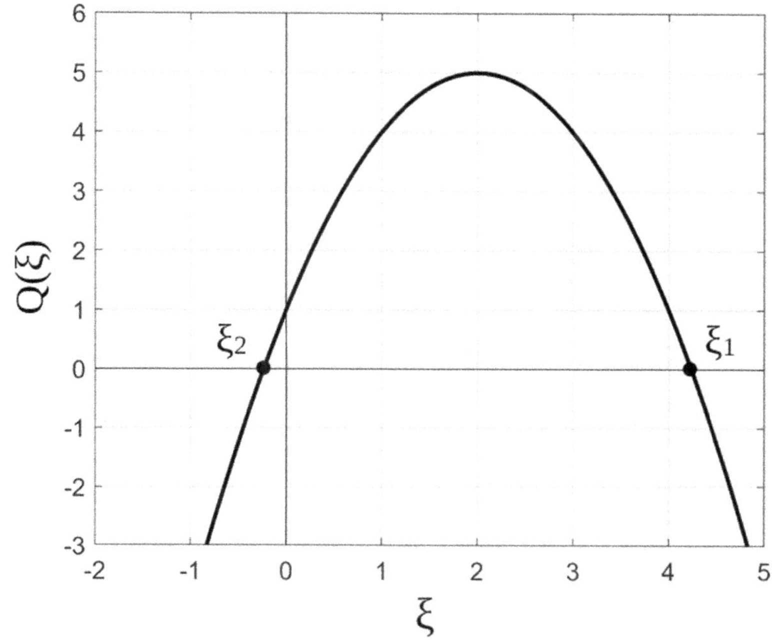

Fig. 5.2 The fundamental quadratics

where $\phi(\lambda) > 0$ satisfies the existence condition of the quadratic function analyzed above and guarantees the convergence of the integral in Eq. 5.47. Finally, combining the complementary solution and the particular solution, we have the *general* solution:

$$F(x) = \frac{1}{\phi(\lambda)}e^{\lambda x} + A_1 e^{\lambda_1 x} + A_2 e^{\lambda_2 x}, \tag{5.61}$$

where it remains to determine the value of the constants.

5.4.2 Geometric Brownian Motion

Suppose now that the problem is to find the solution of

$$G(x) = E \int_0^\infty \varphi(x) e^{-rt} dt, \tag{5.62}$$

where x now follows *geometric Brownian motion*,

$$dx = axdt + \sigma xdZ. \tag{5.63}$$

As before, the *HJB equation* is

$$rG(x)dt = \varphi(x)dt + E\,(dG)\,. \tag{5.64}$$

By applying Ito's lemma, we get

$$E\left(dG\right) = axG^{'}(x)dt + \frac{1}{2}\left(\sigma x\right)^2 G^{''}(x)dt, \tag{5.65}$$

which can be rewritten as

$$\frac{1}{2}\sigma^2 x^2 G^{''}(x) + axG^{'}(x) - rG(x) + \varphi(x) = 0. \tag{5.66}$$

The General Solution

As in the previous section, to find the *complementary* solution of Eq. 5.66, we assume the functional form $G(x) = Ax^{\lambda}$. We get

$$\frac{1}{2}\lambda\left(\lambda - 1\right)\sigma^2 + a\lambda - r = 0 \tag{5.67}$$

$$-\frac{1}{2}\sigma^2\lambda^2 - \left(\frac{1}{2}\sigma^2 - a\right)\lambda + r = 0. \tag{5.68}$$

The new quadratic function has the same properties as the one already analyzed. Therefore, denoting by $\lambda_1 > 0$ and $\lambda_2 < 0$ the two roots of the parabola, we have

$$G(x) = A_1 x^{\lambda_1 > 0} + A_2 x^{\lambda_2 < 0}. \tag{5.69}$$

For the *particular* solution we try instead $\varphi(x) = x^{\lambda}$ and $G(x) = Kx^{\lambda}$. Substituting the latter into Eq. 5.66, we get

$$K\left[\frac{1}{2}\lambda\left(\lambda - 1\right)\sigma^2 + \alpha\lambda - r\right]x^{\lambda} + x^{\lambda} = 0; \tag{5.70}$$

that is,

$$K = \frac{1}{-\frac{1}{2}\lambda\left(\lambda - 1\right)\sigma^2 - \alpha\lambda + r} = \frac{1}{\omega(\lambda)}. \tag{5.71}$$

Therefore, the *general* solution of the problem with geometric Brownian motion is

$$G(x) = \frac{1}{\omega(\lambda)}x^{\lambda} + A_1 x^{\lambda_1 > 0} + A_2 x^{\lambda_2 < 0}, \tag{5.72}$$

with $\omega(\lambda) > 0$. Again, the *boundary conditions* determine the value of the constants A_1 and A_2.

Example 1.
Suppose $G(x)$ is the value of a financial portfolio and x is the *market price* of a *risky asset*. The stochastic dynamic of x is described by Eq. 5.63. What is the expected present value of the portfolio $G(x)$?

To calculate its expected value, we need to solve Eq. 5.72, considering that A_1 and A_2 are two unknowns. We begin by noting that for $x \to 0$, the component x^{λ_2} tends to infinity because $\lambda_2 < 0$, implying that the value of the portfolio increases as the *price* decreases. This is obviously inconsistent with reality since the value of any portfolio tends to decrease with the depreciation of its underlying assets. Similarly, $G(x)$ tends to infinity as x increases because of the component x^{λ_1}, where $\lambda_1 > 0$. This scenario also describes the divergent dynamics of speculative bubbles. Therefore, to ensure that, as x decreases (increases), the function $G(x)$ does not diverge from the particular solution, we must set the coefficients equal to zero, so $A_1 = A_2 = 0$. Finally, the initial condition $x(0) = x_0$ at time $t = 0$ allows us to derive the solution of the problem

$$G(x) = \frac{x_0^{\lambda}}{r - \alpha\lambda - \frac{1}{2}\lambda(\lambda - 1)\sigma^2}, \tag{5.73}$$

with $\omega(\lambda) = r - \alpha\lambda - \frac{1}{2}\lambda(\lambda - 1)\sigma^2 > 0$ to guarantee convergence of the integral.

5.5 Stochastic Optimal Control

So far, we have studied *stochastic dynamic programming* assuming that the objective function is conditioned only by the evolution of the state variable, x. This technique can be generalized to account for more than one stochastic process, x_i, and one or more control variables, z_i, that influence the objective function $f(z_i, x_i)$ with $i = 1, 2, \ldots, n$. In this new scenario, we are confronted with the *optimal stochastic control* problem:

$$\max_{z_1,\ldots z_n} E \int_0^{\infty} f(z_i, x_i)e^{-rt}dt. \tag{5.74}$$

Here, z_i is the control variable chosen to maximize (or minimize) Eq. 5.74, given the Brownian motion, x_i.

Many economic problems take this form. This is the case of stochastic optimal consumption in continuous time, of the investment planning problem, or policy choices for environmental sustainability. These are all problems studied in previous chapters, but now, in their stochastic reformulation, the solution requires the application of Ito's lemma. We next look in detail at five different applications.

5.5.1 Consumption and Portfolio Rules

We start with the budget constraint. Suppose that there are n risky *assets*, the spot price for which follows the geometric Brownian motion

$$dP_i = a_i P_i dt + \sigma_i P_i dz_i \qquad \text{with } i = 1, 2, \ldots, n. \tag{5.75}$$

Suppose that, at time t, an investor owns N_i units of asset i. His or her initial wealth is then

$$W(t) = \sum_{i=1}^{n} N_i(t) P_i(i). \tag{5.76}$$

If, in the interval $(t, t+dt)$, the composition of the portfolio does not change, but the prices of its assets do, the change in financial wealth is

$$dW = \sum_{i=1}^{n} N_i(t) dP_i(i). \tag{5.77}$$

If, finally, in the same interval, consumption $c(t)$ remains constant, the change in net financial wealth will be

$$dW = \sum_{i=1}^{n} N_i(t) dP_i(i) - c(t)dt. \tag{5.78}$$

We denote by $s_i = N_i(t)P_i(i)/W(t)$, the share of wealth invested in a single risky asset i, with $\sum_{i=1}^{n} s_i(t) = 1$. The financial constraint Eq. 5.78 can be rewritten as

$$dW = \sum_{i=1}^{n} s_i(t) a_i W_i(t) dt - c(t)dt + \sum_{i=1}^{n} s_i(t) \sigma_i W_i(t) dz_i, \tag{5.79}$$

which, in the simple case where $n = 1$ (a single risky activity), reduces to

$$dW = (aWdt - c)\, dt + \sigma W dz. \tag{5.80}$$

To close the model, we further assume that the n-th asset n is risk free, that is, $\sigma_n = 0$, with a return equal to r. The constraint becomes

$$dW = \sum_{i=1}^{n-1} s_i(t)\,(a_i - r)\, W_i(t)dt - [rW(t) - c(t)]\, dt + \sum_{i=1}^{n-1} s_i(t) \sigma_i W_i(t) dz_i. \tag{5.81}$$

To describe the optimal stochastic choice let us denote by $u(c(t))$ the consumption function and define $J(W)$ as follows:

$$J(W) = \max_{c(t),\, s(t)} E \int_0^{\infty} u\,[c(t)]\, e^{-rt} dt \quad \text{with} \quad W(0) = W_0 > 0, \tag{5.82}$$

which is the value function for which $c(t)$ and $s(t)$ are the control variables, given the constraint in Eq. 5.81.

Finally, Eq. 5.82 can be rewritten using the HJB equation

$$rJ(W)dt = \max_{c(t),\, s(t)} \{u(c)dt + E\,[dJ(W)]\}\,. \tag{5.83}$$

Using *Ito's lemma*, it becomes

$$rJ(W) = \max_{c(t),\, s(t)} \left\{ \begin{array}{l} u(c) + \left[\sum\limits_{i=1}^{n-1} s_i\,(a_i - r)\,W_i - [rW(t) - c(t)] \right] J^{'}(W) + \\ \qquad + \left[\frac{1}{2} \sum\limits_{i=1}^{n-1} s_i^2 \sigma_i^2 W_i^2 \right] J^{''}(W) \end{array} \right\}. \tag{5.84}$$

The first-order conditions are as follows:

$$c(t) : u^{'}(c) = J^{'}(W). \tag{5.85}$$

$$s(t) : (a_i - r)\,W_i J^{'}(W) + s_i \sigma_i^2 W_i^2 J^{''}(W) = 0. \tag{5.86}$$

The first condition, Eq. 5.85, is the standard Euler equation. Instead, the second condition Eq. 5.86 determines the optimal share s_i of each risky asset held in the optimal portfolio, that is,

$$s_i = \frac{a_i - r}{\sigma_i^2} \left[-\frac{J^{'}(W)}{WJ^{''}(W)} \right]. \tag{5.87}$$

The term in square brackets is the reciprocal of the Arrow-Pratt relative risk aversion index, while the first term is a measure of the risk premium of asset i, as measured by the excess of risky return $(a_i - r)$ over the risk-free asset standardized to its variance σ_i^2. This ratio is known in the literature as *Sharpe's index*.

Solution with a CRRA Function. To calculate the optimal share s_i^* we assume that the consumer has the following CRRA consumption function: $u(c) = \frac{1}{\delta} c^{\delta}$. The corresponding value function $J(W)$ of the *HJB equation* is $J(W) = \alpha + \frac{1}{1-\gamma} W^{1-\gamma}$. Thus, we can write $-\frac{J^{'}(W)}{WJ^{''}(W)} = \frac{1}{\gamma} \frac{W^{-\gamma}}{W^{-\gamma-1}} \frac{1}{W} = \frac{1}{\gamma}$, which implies that

$$s_i^* = \frac{a_i - r}{\sigma_i^2} \frac{1}{\gamma}. \tag{5.88}$$

It follows that the optimal share of the risky asset i decreases with the degree of risk aversion, γ, increases with the expected return a_i, and decreases with the degree of risk σ_i^2.

Solution with the Quadratic Function. In this case, we assume that the consumption function is quadratic, $u(c) = c - \frac{\theta}{2}c^2$. The corresponding value function $J(W)$ is now $J(W) = \alpha + \beta W - \frac{1}{2}\delta W^2$. Hence, we obtain $-\frac{J^{'}(W)}{WJ^{''}(W)} = \frac{\beta - \delta W}{\delta W}$, that is,

$$s_i^* = \frac{a_i - r}{\sigma_i^2}\left(\frac{\beta}{\delta W} - 1\right), \tag{5.89}$$

where $W < \beta/\delta$, so that $J^{'}(W) > 0$.

5.5.2 Consumption and Nonrenewable Resources

Previously, we studied *Hotelling's rule* that captures the relationship between consumption and nonrenewable resources (Chap. 4, Sect. 4.3.1). We now extend this model to the case of uncertainty in continuous time. To simplify the problem, we focus only on intertemporal consumption without considering firm choices.

To calculate the optimal extraction rate of nonrenewable resources in continuous time, we apply the HJB equation and Ito's lemma.

Suppose an oil field is depleted at the rate

$$dV = -qVdt + \sigma V dz, \tag{5.90}$$

where q is the extraction rate and σ the non-voluntary change in the rate. If consumption is $c = qV$, then the problem of optimal consumption of nonrenewable resources can be formulated as follows:

$$J(V) = \max_{q(t)} E \int_0^\infty u(qV) e^{-rt} dt, \tag{5.91}$$

where q is the control variable, with $V(0) = V_0$. The HJB of Eq. 5.91 is

$$rJ(V)dt = \max_q \{u(qV)dt + E[dJ(V)]\}, \tag{5.92}$$

which, by Ito's lemma, becomes

$$rJ(V) = \max_q \left[u(qV) - qVJ^{'}(V) + \frac{\sigma^2}{2}V^2 J^{''}(V)\right], \tag{5.93}$$

the optimal condition for q is

$$u^{'}(qV) = J^{'}(V). \tag{5.94}$$

We obtain an explicit solution to the problem by assuming that the utility is the CRRA $u(c) = \frac{C^{1-\alpha}}{1-\alpha}$, with the corresponding *value function* $J(V) = A + \frac{aV^{1-\alpha}}{1-\alpha}$. The first-order condition is

$$a = q^{-\alpha}. \tag{5.95}$$

Finally, by substituting the two functions and their respective derivatives into the *HJB equation*, we obtain

$$q = \frac{r}{\alpha} + \frac{\sigma^2}{2}(1-\alpha), \tag{5.96}$$

which implies that the optimal choice involves a constant rate of extraction q. Note that if $\sigma^2 = 0$, the rate of extraction reduces to $q = \frac{r}{\alpha}$. Also, it is interesting to note that q increases with uncertainty σ, since $\partial q/\partial \sigma^2 > 0$.

5.5.3 Consumption of Durables

Similar to the discrete-time example in Appendix 2 of Chap. 2, let us now consider a consumer intending to maximize his or her expected utility, which depends on both *durable* and *non-durable* goods.

The consumer's utility is captured by the function

$$u_t = \theta_t \left(c_t^{\alpha} D_t^{1-\alpha}\right), \tag{5.97}$$

where c_t is the amount of *non-durable goods* and D_t the stock of *durable ones*. In addition, consumption is subject to the shock θ_t capturing risk in the goods market. The purchase and use of durables takes place over time, and in each period, the consumer buys an amount of durables F_t, the price of which is p^D, bearing the additional cost $C(F) = \frac{\gamma}{2}F_t^2$ caused by transaction costs. The stock of durables evolves according to

$$dD_t = (F_t - \delta D_t)dt. \tag{5.98}$$

Equation 5.98 states that the consumption of durables increases over time at the gross rate F_t, but decreases at the rate δD_t due to their depreciation. Finally, the equation

$$d\theta_t = \sigma\theta_t dz_t \tag{5.99}$$

closes the model. It describes the behavior of the shock where dz_t is a Wiener process.

Assuming that the discount rate ρ is constant, we can write the expected value function as

$$J(D_t, \theta_t) = \max_{F_s, c_s} E_t \int_t^{\infty} \left[\theta_s c_s^{\alpha} D_s^{1-\alpha} - p^c c_s - p^D F_s - \frac{\gamma}{2}{F_s}^2\right] e^{-\rho(s-t)} ds, \tag{5.100}$$

where p^c is the cost of non-durable goods and p^D the one of durables.

The maximization of consumption is subject to the constraints in Eqs. 5.98 and 5.99. The corresponding HJB equation is

$$\rho J(D_t, \theta_t)dt = \max_{F_t, c_t} \left[\theta_t c_t^{\alpha} D_t^{1-\alpha} - p^c c_t - p^D F_t - \frac{\gamma}{2} F_t{}^2\right] dt + E_t(dV). \tag{5.101}$$

We compute the expected value, $E\ (dV)$, by applying Ito's lemma:

$$dV = J_D(dD) + V_\theta(d\theta) + \frac{1}{2} V_{DD}(dD)^2 + \frac{1}{2} J_{\theta\theta}(d\theta)^2. \tag{5.102}$$

Substituting Eqs. 5.98 and 5.99 into Eq. 5.102, and recognizing that $E_t(dz) = 0$, we get the expected utility over the time interval dt

$$E_t(dV) = \left[(F_t - \delta D_t) J_D + \frac{1}{2} \theta_t^2 \sigma^2 J_{\theta\theta}\right] dt. \tag{5.103}$$

Substituting Eq. 5.103 into Eq. 5.101 yields

$$\rho J(D_t, \theta_t) = \max_{F_t, c_t} \left\{\theta_t c_t^{\alpha} D_t^{1-\alpha} - p^c c_t - p^D F - \frac{\gamma}{2} F^2 \right. \\ \left. + (F_t - \delta D_t) J_D + \frac{1}{2} \theta_t^2 \sigma^2 J_{\theta\theta}. \right. \tag{5.104}$$

To solve the model, we rely on a two-step strategy (Abel 1983). At first, the consumer chooses c_t to maximize the instantaneous utility from *current* consumption

$$\max_{c_t} \left[\theta_t c_t^{\alpha} D_t^{1-\alpha} - p^c c_t\right] = m \theta_t{}^{\beta} D_t, \tag{5.105}$$

where

$$m = (1-\alpha) \left(\frac{\alpha}{p^c}\right)^{\frac{\alpha}{1-\alpha}} \quad \text{and} \quad \beta = \frac{1}{1-\alpha}. \tag{5.106}$$

Note that in Eq. 5.105 $m\theta_t{}^{\beta}$ represents the marginal utility of durables at time t. Differentiating the right-hand side of Eq. 5.104 with respect to F_t, we get the optimal marginal utility of durable goods

$$J_D = \gamma F_t + p^D. \tag{5.107}$$

Substituting Eqs. 5.105 and 5.107 into the HJB in Eq. 5.104, we obtain the equation

$$\rho J(D_t, \theta_t) = m \theta_t{}^{\beta} D_t + \frac{\gamma}{2} F_t{}^2 - \delta D_t J_D + \frac{1}{2} \theta_t^2 \sigma^2 J_{\theta\theta}. \tag{5.108}$$

Equations 5.108 and 5.107 can be expressed together as a nonlinear *second-order partial differential equation*. Such equations are often difficult to solve. However, in this example, a closed-form solution exists (Saltari and Travaglini 2011). We assume that $J(D_t, \theta_t)$ is a linear function of D_t so that

$$J(D_t, \theta_t) = J_D(\theta_t)D_t + G(\theta_t), \tag{5.109}$$

where

$$J_D = \frac{m\theta^\beta}{\rho + \delta - \frac{\alpha(\beta\sigma)^2}{2}} \tag{5.110}$$

and

$$G(\theta) = \frac{1}{2\gamma}\left[\frac{J_D^2}{\rho - (1+\alpha)(\beta\sigma)^2} - \frac{2J_D p^D}{\rho - \frac{\alpha(\beta\sigma)^2}{2}} + \frac{\left(p^D\right)^2}{\rho}\right], \tag{5.111}$$

with

$$F_t = \frac{1}{\gamma}\left[J_D - p^D\right]. \tag{5.112}$$

It can be easily verified that the equations above satisfy both Eqs. 5.107 and 5.108. Finally, note that from Eq. 5.110) follows that an increase in σ^2 leads to a corresponding increase in the optimal rate of consumption of durables since

$$\frac{\partial F_t}{\partial \sigma^2} = \frac{\alpha\,\beta^2\,m\,\theta^\beta}{2\,\gamma\left(-\frac{\alpha\,œ\,\beta^2}{2} + \delta + \rho\right)^2} > 0. \tag{5.113}$$

This is a *direct* effect: a greater uncertainty encourages consumers to buy more durables. However, the *global effect* of a higher uncertainty on durables concerns the overall value function $J(D_t, \theta_t)$. This effect is ambiguous since a rising σ^2 also affects the value $G(\theta)$ of the durables stock.

5.5.4 Consumption and Growth

Let us assume that the consumption function is the CRRA:

$$u(c) = \frac{c^{1-\alpha}}{1-\alpha} \tag{5.114}$$

and that the production function is the Cobb-Douglas

$$f(k) = Ak^\alpha, \tag{5.115}$$

where $c = \frac{C}{L}$ and $k = \frac{K}{L}$, and where we assume, for simplicity, that the elasticity of the production function is equal to the degree of risk aversion. Recall that if the population growth rate follows a geometric Brownian motion, capital accumulation is described by the equation

$$dk = \left[f(k) - c - \left(n + \delta - \sigma^2\right) k\right] dt - \sigma^2 k dz. \tag{5.116}$$

The *stochastic* Ramsey problem can thus be formulated as

$$\begin{cases} & J(k) = \max_{c(t)} E \int_0^\infty u(c_t) e^{-\rho t} dt \\ \text{s.t.} & dk = \left[f(k) - c - \left(n + \delta - \sigma^2\right) k\right] dt - \sigma^2 k dz \\ & k(0) = k_0 \text{ e } k_t, c_t \geqslant 0. \end{cases} \tag{5.117}$$

The corresponding HJB equation is

$$\rho J(k) dt = \max_c \{u(c) dt + E[dJ(k)]\} \tag{5.118}$$

$$\rho J(k) = \max_c \left\{\frac{c^{1-\alpha}}{1-\alpha} + \left[Ak^\alpha - c - \left(n + \delta - \sigma^2\right) k\right] J'(k) + \frac{1}{2}\sigma^2 k^2 J''(k)\right\}. \tag{5.119}$$

The first-order condition is

$$c^{-\alpha} = J'(k) \tag{5.120}$$

and the HJB equation can be rewritten as

$$0 = \frac{\alpha}{1-\alpha}\left[J'(k)\right]^{-\frac{1-\alpha}{\alpha}} - \rho J(k) + \left[Ak^\alpha - \left(n + \delta - \sigma^2\right) k\right] J'(k) + \frac{1}{2}\sigma^2 k^2 J''(k). \tag{5.121}$$

If the production function $f(k) = Ak^\alpha$ does not appear in Eq. 5.121, the latter could be solved using the solution $a^{-1}k^{1-\alpha}$. But, the presence of the production function requires an additional restriction to solve the problem. In particular, we assume that the term $(Ak^\alpha) J'(k)$ is a constant. With this assumption, the *value function* takes the form

$$J(k) = (1-\alpha)^{-1} a^{-1} k^{1-\alpha} + b, \tag{5.122}$$

so that the first-order condition becomes

$$c = a^{\frac{1}{\alpha}} k \tag{5.123}$$

and consumption is a linear function of the capital-to-labor ratio.

Finally, since $J'(k) = a^{-1}k^{-\alpha}$ and $J''(k) = -\alpha a^{-1} k^{-(1+\alpha)}$, by substituting in the HJB and simplifying we get

$$a = \left\{\frac{1-\alpha}{\alpha}\left(n + \delta - \sigma^2 + \frac{\sigma^2}{2}\alpha\right) + \frac{\rho}{\alpha}\right\}^\alpha \tag{5.124}$$

and

$$b = \rho^{-1} A a^{-1}. \tag{5.125}$$

As mentioned above, this closed-form solution is a special case, which is, however, not compatible with even small changes in the model. For example, if the production function is still a Cobb-Douglas, but the elasticity of production is different from the degree of consumer risk aversion, the previous value function *cannot be* the solution to the problem in Eq. 5.117.

Correspondence between HJB equation and Hamiltonian

Note that in the absence of uncertainty, the HJB equation of the previous problem reduces to

$$\rho J(k) = \max_c \left\{ u(c) + [f(k) - c - (n+\delta)k] J^{'}(k) \right\}, \tag{5.126}$$

the first-order condition of which is

$$u^{'}(c) = J^{'}(k), \tag{5.127}$$

and the envelope gives rise to the relation

$$\rho J^{'}(k) = \left[f^{'}(k) - (n+\delta) \right] J^{'}(k) + [f(k) - c - (n+\delta)k] J^{''}(k). \tag{5.128}$$

It is easy to verify that the derivative with respect to time of the first-order condition is

$$u^{''}(c)\dot{c} = J^{''}(k)\dot{k}; \tag{5.129}$$

that is, $\frac{u^{''}(c)\dot{c}}{\dot{k}} = J^{''}(k)$, where $\dot{k} = \frac{dk}{dt}$ and $\dot{c} = \frac{dc}{dt}$. Substituting both these expressions and the first-order condition in the envelope equation, we get

$$\rho u^{'}(c) = \left[f^{'}(k) - (n+\delta) \right] u^{'}(c) + [f(k) - c - (n+\delta)k] \frac{u^{''}(c)\dot{c}}{\dot{k}}, \tag{5.130}$$

from which, by simplifying, we get

$$\dot{c}_t = -\frac{u^{'}(c_t)}{u^{''}(c_t)} \left[f^{'}(k_t) - (n+\delta+\rho), \right] \tag{5.131}$$

which is the optimal consumption rule of Eq. 4.23 of the Ramsey problem in Chap. 4, Sect. 4.1.4.

5.5.5 Regulated Brownian Motion

As shown in Fig. 4.4, a Brownian motion can fluctuate randomly around a deterministic trend with a range over the entire real line $(-\infty, +\infty)$. However, there are often restrictions on the range. The literature names these stochastic processes *regulated* Brownian motions. In probability theory, a regulated Brownian motion is a Wiener process that evolves in a space with reflective or absorbing barriers. In economics, these regulated diffusion motions are commonly used to approximate the phenomena related to the presence of price barriers, exchange rate ceilings, market regulations, and even liquidity constraints and credit rationing (Froot and Obstfeld 1989; Krugman 1991, Bertola 1998; Saltari and Travaglini 2001).

In what follows, we present an example of stochastic consumption and liquidity constraints in continuous time. Specific boundary conditions are necessary to solve the optimal consumption problem.

5.5.6 Stochastic Income and Constrained Consumption

Let us assume that households derive their expected utility from consumption and that the time horizon is infinite. The maximization problem is then

$$\begin{cases} \max\limits_{c_s} \int_t^\infty E_t\left[u\left(c_s\right)\right] e^{-r(s-t)} ds \\ \text{s.t.} \quad \dot{a}_s = r a_s + y_s - c_s, \end{cases} \tag{5.132}$$

where c_s and y_s are consumption and income, a_s is wealth, and $E_t\left[\cdot\right]$ denotes expectations at time t. We assume that at the initial time, a_t is given. The transversality condition is $\lim\limits_{s\to\infty} a_s e^{-r(s-t)} = 0$.

As shown in Sect. 2.2.3, if the utility function is quadratic, then $E_t\left(c_s\right) = c_t$, for all $s \geq t$. Together with the solution of the budget constraint Eq. 5.132, it provides the following consumption function:

$$c_t = \frac{r}{1+r}\left[a_t + h_t\left(y_t\right)\right], \tag{5.133}$$

where

$$h_t\left(y_t\right) = \int_t^\infty E_t\left(y_s\right) e^{-r(s-t)} ds \tag{5.134}$$

is the discounted present value of expected future income at time t. With uncertainty, the income path is stochastic, depending on both the growth rate μ and volatility σ. Suppose that the geometric Brownian motion

$$dy_t = \mu y_t dt + \sigma y_t dz \tag{5.135}$$

describes this dynamic. Also, we assume that the household is credit rationed so that it can rely only on current and future income to consume. Following Besley (1995) we can write

$$0 \leq y_t \leq y_{\max}. \tag{5.136}$$

This is the liquidity constraint: it states that income cannot fall below the lower level $y_{min} = 0$ (minimum wage) and cannot rise beyond the upper barrier y_{max} (maximum wage).

This consumption problem admits a closed-form solution. Notice that $h_t(y_t)$ is a function of the geometric Brownian motion y_t. Thus, to calculate the changes of h_t, we must apply Ito's lemma to compute the differentials of $h_t(y_t)$ with respect to y_t. We get

$$\frac{1}{dt}E_t(dh) = \mu y h'(y_t) + \frac{1}{2}h''(y_t)y_t^2\sigma^2, \tag{5.137}$$

where $h'(y_t)$ and $h''(y_t)$ denote the partial derivatives of $h(y_t)$ with respect to y_t. Further, recall that Eq. 5.134 can be rewritten as (see Chap. 5, Sect. 5.4.1)

$$rh_t = y_t + E_t\left(\frac{dh}{dt}\right), \tag{5.138}$$

which is the well-known HJB equation. Substituting for the expectation in Eq. 5.138, we obtain

$$\frac{1}{2}h''(y_t)\sigma^2 y_t^2 + h'(y_t)\mu y_t + y_t - rh(y_t) = 0. \tag{5.139}$$

This is a second-order differential equation, the solution of which describes how the discounted value of future income changes when y_t fluctuates randomly in the range $0 \leq y_t \leq y_{\max}$.

The *complementary* function of Eq. 5.139 is a linear combination of any two linearly independent solutions, say

$$h(y_t) = B_1 y_t^{\alpha_1} + B_2 y_t^{\alpha_2}, \tag{5.140}$$

where B_1 and B_2 are two constants to be determined from the boundary conditions, and $\alpha_1 > 0$ and $\alpha_2 < 0$ are the roots of the quadratic equation

$$\frac{1}{2}\alpha^2\sigma^2 + \left(\mu - \frac{\sigma^2}{2}\right)\alpha - r = 0. \tag{5.141}$$

If y_t is free to range over the entire line $(-\infty, +\infty)$, we know that only the *particular* solution of the differential equation characterizes the general solution of the problem. But, if y_t is restricted, then we can get some information by considering what happens to the extremes of the range. The constraint Eq. 5.136 implies that $B_2 = 0$. Indeed, since income cannot be smaller than zero, we must set $B_2 = 0$ to prevent consumption from becoming larger and larger as income tends to zero.

Similarly, if y_t has an upper barrier at $y_{\max}$, then the inspection of $h(y)$, as y_t goes to infinity, suggests that some additional condition must be added to compute the value of B_1 when the constraint binds in order to assure the respect of the Euler equation at the barrier.

To solve this latter problem, we need the additional information provided by the *particular* solution of the full Eq. 5.139. This particular solution must be guessed. We consider a function of the form $h(y) = Ay_t$. Substituting in Eq. 5.139, we get the value of the constant A, that is, $A = 1/(r - \mu)$.

Therefore, the general solution of Eq. 5.139 is

$$h(y_t) = y_t/(r - \mu) + B_1 y_t^{\alpha_1}, \tag{5.142}$$

where

$$\alpha_1 = \frac{\frac{\sigma^2}{2} - \mu + \sqrt{\left(\mu - \frac{\sigma^2}{2}\right)^2 + 2r\sigma^2}}{\sigma^2} > 0 \tag{5.143}$$

and where B_1 remains to be determined.

Constrained consumption

We need a rule in which consumption is a function $c_t = c\,(y_t)$ to find the optimal consumption policy, describing the *switching* path between the unconstrained and the constrained status. Substituting Eq. 5.142 in Eq. 5.133, and assuming (to simplify the problem) that $a_t = 0$, we get the consumption function

$$c(y_t) = \frac{r}{(1+r)\,(r - \mu)} y_t + \frac{r}{(1+r)} B_1 y_t^{\alpha_1}. \tag{5.144}$$

where $0 < \frac{r}{(1+r)(r-\mu)} < 1$ is the marginal propensity to consume. Once again, if the upper barrier is absent, even the constant B_1 must be set equal to zero because *no liquidity constraint* occurs when $y_t \to \infty$.

If the constraint limits the random income process, the upper barrier impinges upon current consumption. In this case, B_1 exerts its effects on consumption even in the intermediate phases before the constraint binds. Indeed, as long as y_t lies within the range, its evolution is described by Eq. 5.135. Once y_t reaches y_{max}, its process changes since income can only decrease randomly. Thus, the level of consumption approaches its maximum value c_{max} when income approaches the threshold y_{max}. For the same reason, consumption decreases when income drops.

Any forward-looking household anticipates this process and finds it optimal to select a consumption policy that accounts for future constraints at the outset. To maximize the expected utility, the household *smooths* consumption along the entire path, even in the neighborhood of y_{max}. If this condition were to be violated, the consumption function $c(y_t)$ would have no derivative in y_{max}, implying a violation

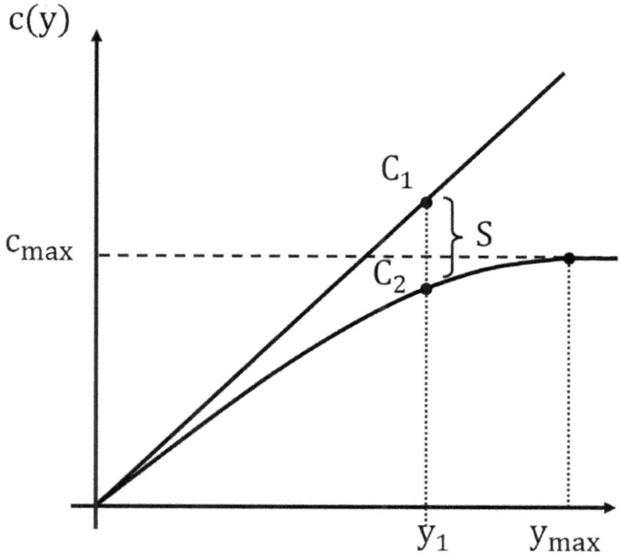

Fig. 5.3 Consumption function with an upper bound

of the Euler equation along the optimal switching path (see Saltari and Travaglini 2006). The boundary condition corresponding to this control is

$$c'(y_{max}) = 0. \tag{5.145}$$

Usually, this condition, which pertains to the first-order derivative of an expected present value of a function of Brownian motion, is called the *smooth pasting condition* (Dixit 1993). It is sufficient to fix the constant B_1. By applying Eq. 5.145 to Eq. 5.144 we get

$$c(y_t) = \frac{r}{(1+r)(r-\mu)}\left[y_t - \frac{y_t^{\alpha_1}}{r\alpha_1 (y_{\max})^{\alpha_1 - 1}}\right]. \tag{5.146}$$

Figure 5.3 shows the linear function y_t and the locus representing the nonlinear functional form $c(y_t)$ which is tangent at c_{max} when the liquidity constraint binds. Note that without upper barriers, the level of consumption would rise along the straight path y_t. With liquidity constraint, instead, the consumption function becomes *concave* smoothly since the household anticipates the effect of the future constraint: the household perceives closeness to the upper barrier as an exacerbation of the liquidity constraint, becoming more reluctant to consume. There are two consequences of this behavior.

First, note that *saving* increases over the range of y_t. Thus, although the utility function is quadratic, the liquidity constraint affects the consumption policy even in the intermediate periods when it is still slack. Moreover, the optimal intertemporal consumption depends not only on the expected value μ of income, as in the certainty equivalence model, but also on the variance σ^2, which determines the value of the positive root $\alpha_1 > 0$ and the curvature of the consumption function. No less important is the fact that c_{max} is an endogenous value that depends on the parameters of the problem, including the σ^2: when the variance increases, c_{max} decreases while saving raises. This *extra* quota of saving can be seen as *precautionary saving*. Second, as noted, along the optimal *switching* path, consumption increases more slowly than income, so saving also increases. Figure 5.3 illustrates what happens when income is equal to y_1: if a household behaves in a *myopic* way, consumption would be C_1 on the straight line. However, this is not possible because of the upper barrier. Thus, the forward-looking household anticipates that income will never exceed y_{max} and that the closer the liquidity constraint, the higher the probability that income and consumption will be lower in the future (Deaton 1991, 1992). This bearish expectation affects consumption in any period, even when the constraint is slack, shifting down consumption along the concave curve in C_2. The distance between this curve and the straight line provides a measure of the extra *saving* (S). Further, consumption tends to change as the variance changes. From Eq. (5.146), it follows that $\frac{dc}{d\sigma^2} < 0$. This is the result of two effects. On the one hand, a higher σ^2 decreases the maximum consumption associated with the upper barrier y_{max}. On the other hand, it also decreases the level of consumption in each period along the convex curve describing the *switching path*, leading, optimally, to the household's consumption moving from the unconstrained to the constrained status. To conclude, in the presence of liquidity constraints, the propensity to consume is much greater at low than at high levels of income, and consumption smoothing implies increasing *precautionary saving* as income and consumption approach their maximum level. Importantly, this happens *even* in the case of a quadratic utility function.

Further Reading
A rigorous (yet friendly) introduction to the theory of stochastic integration is in Cohen and Elliott (2015) (Parts II to IV). For stochastic optimization in continuous time applied to economic problems, the reference textbook is Chang (2004). Deterministic and stochastic optimal control is well discussed in Fleming and Rishel (2012). Continuous-time Markov processes are explored rigorously in Fleming and Soner (2006), and Brownian motion in Karatzas and Shreve (1991). Other notable textbooks, especially for the applications presented, are *Stochastic Methods in Economics and Finance* by Malliaris and Brock (1977); *Investment Under Uncertainty* by Dixit (1993), *The Art of Smooth Pasting* by Dixit (1993); *Dynamic Asset Pricing Theory* by Duffie (1992); *Optimal Control Theory. Applications to Management Science and Economics* by Sethi (2019); and *The Economics of Inaction. Stochastic Control Models with Fixed Costs* by Stokey (2008).

References

Abel, A. B. (1983). Optimal investment under uncertainty. *The American Economic Review, 73* (1), 228–233.

Bertola, G. (1998). Irreversible investment. *Research in Economics, 52* (1), 3–37.

Besley, T. (1995). Property rights and investment incentives: Theory and evidence from ghana. *Journal of political Economy, 103* (5), 903–937.

Calcagnini, G., & Saltari, E. (2000). Real and financial uncertainty and investment decisions. *Journal of Macroeconomics, 22* (3), 491–514.
Chang, F.-R. (2004). *Stochastic optimization in continuous time*. Cambridge University Press.
Cohen, S. N., & Elliott, R. J. (2015). *Stochastic calculus and applications*. Springer.
Deaton, A. (1991). Saving and liquidity constraints. *Econometrica, 59*(4), 1221–1248.
Deaton, A. (1992). *Understanding consumption*. Oxford University Press.
Dixit, A. K. (1993). *The art of smooth pasting. First edition*. Taylor & Francis.
Duffie, D. (1992). *Dynamic asset pricing theory*. Princeton University Press.
Fleming, W. H., & Rishel, R. W. (2012). *Deterministic and stochastic optimal control* (Vol. 1). Springer Science & Business Media.
Fleming, W. H., & Soner, H. M. (2006). *Controlled markov processes and viscosity solutions* (Vol. 25). Springer Science & Business Media.
Froot, K. A., & Obstfeld, M. (1989). Stochastic process switching: Some simple solutions. (2998).
Itô, K. (1944). Stochastic integral. *Proceedings of the Imperial Academy Tokyo, 20* (8), 519–524.
Itô, K. (1951). On a formula concerning stochastic differentials. *Nagoya Mathematical Journal, 3*.
Karatzas, I., & Shreve, S. (1991). *Brownian motion and stochastic calculus*. Springer Science & Business Media.
Krugman, P. R. (1991). Target zones and exchange rate dynamics. *The Quarterly Journal of Economics, 106* (3), 669–682.
Malliaris, A. G., & Brock, W. A. (1977). *Stochastic methods in economics and finance*. North Holland.
Saltari, E., & Travaglini, G. (2001). Financial constraints and investment decisions. *Scottish Journal of Political Economy, 48* (3), 330–344.
Saltari, E., & Travaglini, G. (2006). The effects of future financing con- straints on capital accumulation: Some new results on the constrained investment problem. *Research in Economics, 60* (2), 85–96.
Saltari, E., & Travaglini, G. (2011). Optimal abatement investment and environmental policies under pollution uncertainty. In O. de La Grandville (Ed.), *Frontiers of economic growth and development* (pp. 527–541). Emerald.
Sethi, S. P. (2019). *Optimal control theory applications to management science and economics*. Princeton University Press.
Stokey, N. L. (2008). *The economics of inaction: Stochastic control models with fixed costs*. Princeton University Press.

GPSR Compliance
The European Union's (EU) General Product Safety Regulation (GPSR) is a set of rules that requires consumer products to be safe and our obligations to ensure this.

If you have any concerns about our products, you can contact us on

ProductSafety@springernature.com

In case Publisher is established outside the EU, the EU authorized representative is:

Springer Nature Customer Service Center GmbH
Europaplatz 3
69115 Heidelberg, Germany

www.ingramcontent.com/pod-product-compliance
Ingram Content Group UK Ltd.
Pitfield, Milton Keynes, MK11 3LW, UK
UKHW020843290926
15470UKWH00004B/1520
9783031549854